Fermentation - Processes, Benefits and Risks

Edited by Marta Laranjo

Published in London, United Kingdom

IntechOpen

Supporting open minds since 2005

Fermentation - Processes, Benefits and Risks
http://dx.doi.org/10.5772/intechopen.87513
Edited by Marta Laranjo

Contributors
Romel E. Guzmán-Alvarez, José G. Márquez-Ramos, Michael Osho, Sarafadeen Kareem, Reta Reta, Salengke Salengke, Junaedi Muhidong, Mursalim Mursalim, Sitti Nurmiah, Ophirtus Sumule, Fitri Fitri, Tridjoko Wisnu Murti, Tolulope Oreoluwa Faniyi, Olukayode Stephen Oyatokun, Anouk F. F. Duque, Mónica Carvalheira, Felix Offei, Puneet Kumar Gupta, Jyotheeswara Reddy Edula, Laslo Éva, Éva György

First published in London, United Kingdom, 2021 by IntechOpen
IntechOpen is the global imprint of INTECHOPEN LIMITED, registered in England and Wales, registration number: 11086078, 5 Princes Gate Court, London, SW7 2QJ, United Kingdom
Printed in Croatia

British Library Cataloguing-in-Publication Data
A catalogue record for this book is available from the British Library

Additional hard and PDF copies can be obtained from orders@intechopen.com

Fermentation - Processes, Benefits and Risks
Edited by Marta Laranjo
p. cm.
Print ISBN 978-1-83968-816-4
Online ISBN 978-1-83968-817-1
eBook (PDF) ISBN 978-1-83968-818-8

For EU product safety concerns:
IN TECH d.o.o., Prolaz Marije Krucifikse Kozulić 3, 51000 Rijeka, Croatia, info@intechopen.com or visit our website at intechopen.com.

Meet the editor

Marta Laranjo holds a Ph.D. in Biology and is a senior researcher at MED - Mediterranean Institute for Agriculture, Environment and Development (Food Science and Technology Group), Universidade de Évora, Portugal. She is also the executive director of the UNIMED Subnetwork on Food and Water. An expert on microbiology and molecular biology, her main interests include food microbiology, with a special emphasis on starter cultures and the dynamics of food microbiota. She has published several book chapters and numerous research papers in international peer-reviewed journals. She is also a frequent reviewer for several journals in the areas of microbiology and molecular biology. Dr. Laranjo has supervised several Ph.D., Master, and undergraduate students.

Contents

Preface

Fermentation - Processes, Benefits and Risks deals with the topic of fermentation, a process by which carbohydrates are converted to alcohol or organic acids under anaerobiosis using microorganisms (bacteria or yeasts). It is divided into two sections: "Fermentation Processes in Agriculture and Industry" and "Food Fermentations."

The first section includes four chapters that introduce the role of fermentation in agriculture, as well as the various industrial applications of fermentation.

In Chapter 1, the authors introduce fermentation as a multi-disciplinary concept in agriculture, exploring the fermentative processes in the nutrition system of ruminants and unraveling how fermentation processes occur in the soil. In Chapters 2 to 4, the authors focus on distinct industrial applications of different fermentation processes, including biopharmaceutical (Chapter 2), biocatalysis of enzymes (Chapter 3), and sustainable biomass processing via an integrated biorefinery approach (Chapter 4).

The second section contains five chapters on different aspects of fermented foods, which comprise very diverse food products and beverages with a long history of worldwide importance for human nutrition, health, and the economy. The main groups of fermented foods include cereals (bread and sake); vegetables, fruits, and legumes (table olives, pickles, and vinegar); milk (cheese, yogurt, and kefir); meat and meat products (ham and sausages); fish products (anchovies and herring); and alcoholic beverages (wine and beer). A huge diversity of indigenous microorganisms, mostly bacteria and yeasts, is associated with this enormous variety of fermented foods. They differ not only in the used raw materials, and corresponding final products, but also in their distinct fermentative behavior. The microbial safety of these fermented food products is, however, threatened by the potential presence of foodborne pathogens. These may be harmful per se, but also due to the toxins and other secondary metabolites they release, such as mycotoxins or biogenic amines. To assure the quality and safety of fermented foods, different approaches are needed to successfully control the indigenous microbiota and conduct fermentation, for example, using starter cultures, but also to effectively implement innovative preservation methodologies.

In Chapter 5, the authors review the fermentation of milk from distinct origins, namely bovine, non-bovine, and vegetable, discussing the potential benefits and drawbacks of each of them. Chapter 6 gives us some insights into the microbiota of traditionally processed cheeses, according to their different compositions and production technologies. Chapter 7 deals with the fermentation of cocoa beans, a crucial stage in the post-harvest processing of cocoa that has not been studied extensively. Chapter 8 explores ohmic fermentation of coffee as an innovative approach to imprint a distinctive aroma profile and improve the aroma quality of coffee. In Chapter 9, the authors show how added-value compounds, such as volatile fatty acids, can be produced by biotechnological processes, namely anaerobic fermentation, valorizing food waste as a carbon source for fermentation towards a circular economy.

This book encourages the reduction of food loss and food waste and promotes a circular economy via finding alternative pathways to fully use natural resources through fermentation processes.

The United Nations (UN) Sustainable Development Goals (SDGs) are the world's shared plan to end extreme poverty, reduce inequality, and protect the planet by 2030. This agenda includes seventeen specific SDGs, and the research compiled in this book intends to contribute to at least three of them, namely, SDS 2: End Hunger; SDG 9: Industry, Innovation, and Infrastructure; and SDG 12: Responsible Consumption and Production.

The editor was happy to work on this book and hopes that its content will be useful to both students and researchers, promoting debate on how fermentation processes can contribute to more sustainable production and consumption in the future.

Marta Laranjo
MED-Universidade de Évora,
Évora, Portugal

Section 1

Fermentation Processes in Agriculture and Industry

Chapter 1

Fermentation in the Perspective of Agriculture

Tolulope Oreoluwa Faniyi and Olukayode Stephen Oyatokun

Abstract

Fermentation is a multi-disciplinary concept that is defined from the perspectives of various disciplines. It connotes different meanings to microbiologist, biochemist, food and nutritionist (rumen modulator/manipulation) and soil scientist. However, the overall is that it results in the breakdown of substrates (organic or inorganic) in the absence of air to yield intermediate by-products including methane gas (loss of feed energy). The concept includes activities bothering on anaerobic and aerobic processes to enhance the breakdown of substrates to produce some useful materials and biogases. Although carbohydrates are often regarded as essential materials for fermentations, organic acids (including amino acids), proteins, fats, and other organic compounds are fermentable substrates for selected microorganisms with the production of total volatile fatty acids and their individual components (i.e. propionates, butyrates and acetates). Today, fermentative process involves the transformation of raw materials, aerobically or anaerobically, to other valuable products through the activities of microorganisms.

Keywords: fermentation, substrate, microorganism, breakdown, intermediate products

1. Introduction

Fermentation is a concept in use by many disciplines; hence it is a multidisciplinary concept that is defined from the perspectives of various disciplines involved. It connotes different meanings to scientists working in the areas of microbiology, biochemistry, food nutrition, animal nutrition and soil science. However, the overall is that fermentation is a process that results in the breakdown of substances (organic or inorganic) where there is no air (oxygen) to yield some intermediate by-products including biogas.

Fermentation finds its root in the word "ferment", which is a Latin word *fervere* meaning "to boil". It is a process of natural evolution and people use fermentation to produce wine and alcohol long before the advent of science and the understanding of biochemical processes.

Today, this ancient knowledge in combination with scientific knowledge is applied in production processes to make a variety of products that are useful for life's existence. Hence, fermentation is a metabolic process which alters the chemical composition of organic substrate following the activities of enzymes. Its primary function in microorganism is to produce energy (adenosine triphosphate, ATP) by breaking down of organic nutrients [1]. Humans use fermentation for the production of beverages and foodstuffs as well as for making wines, beers and yoghurts [2].

The concept of fermentation in today's world has grown to include those activities bothering on both anaerobic and aerobic processes that enhance the breakdown of organic and inorganic substances to produce some forms of useful materials and biogases. Although carbohydrates are usually regarded as necessary materials for fermentations, other organic compounds such as proteins, fats, and organic acids are substrates that can be fermented by specialized microorganisms. These materials serve as food and energy sources to microorganisms. Under anaerobic conditions, a fraction of the potential energy is liberated because of incomplete oxidation, leaving a heap of unoxidized organic by-products. In order to obtain as much energy obtainable under aerobic conditions, several molecules of glucose need be broken down under anaerobic conditions. Fermentation can therefore be regarded, as the breakdown of organic compounds, where oxygen is lacking, to organic intermediate products, which the cell's enzyme system could not further oxidize except oxygen is available. The product of fermentation varies as the microorganisms acting on the substrates, depending on the cell's enzymes complex and the conditions of the environment. The economic importance of these by-products marks the advent of industrial microbiology.

The knowledge of fermentation as anaerobic process brought out the difference between microbial biochemistry and biochemistry of mammalian tissues. Fermentative process was postulated to follow similar paths because the intermediate products of glucose metabolism were similar. Consequently, the fermentation of carbohydrates by microorganisms was considered similar to glycolysis by mammals. This explains why many authors employ the terms "glycolysis" for the description of one method of anaerobic breakdown of carbohydrates by microorganisms and why the terms: "fermentation" and "glycolysis" are used synonymously. These two terms are different because glycogen is not stored by bacteria and lactate was not the intermediate or end product during breakdown of carbohydrates by bacteria.

Therefore, the main aim of this chapter is to define fermentation in broader term beyond what is currently used to describe it. The specific objectives are to explore the fermentative processes in the nutrition system of ruminants and showcase how fermentation processes occur in the soil system to give rise to available plant nutrient elements.

2. Fermentation in ruminants

2.1 Ruminants nutrition

Ruminants possess rumen as part of a complex stomach hosting diverse microbial matrix (bacteria, protozoa, fungi and archaea i.e. methanogens) that helps to hasten the breakdown of all the solids (forages, roughages, crop residues, agro-industrial by-products and fibers or feed substrates) and help boost fermentation to enhance and raise productivity [3, 4]. Seasonal feed challenge in ruminants together with low intake and poor digestibility contributes to their low productivity [5, 6]. Also, fermentation that is inefficient reduces the potential satisfaction ruminants derive from the feed consumed.

It was reported that quite a number of chemical feed additives such as antibiotics, ionophores, methane inhibitors, defaunating agents, etc. have been used in ruminant nutrition to manipulate, modify or to improve rumen fermentation and degradability swith the aim of boosting or hastening the rumen efficiency [4, 7–11].

The use of banned artificial antibiotics by European Union (EU) in the diets of livestock to increase production of meat, milk and wool, as well as suppression of some microbial activities has motivated researchers' interest in the use of plants and plant extracts [7–9, 12].

It is imperative now to know that the fate of the underlining facts or factors of ruminant feeding, fiber degradation, digestion and metabolism lies on the ruminal microbial ecology or rumen ecosystem (i.e. bacteria, archaea (methanogens), fungi and protozoa) with their various activities. Ruminants establish a symbiotic relationship with rumen Microorganisms by providing nutrients and optimal environment for fermentation of feeds, degradation of fiber and synthesizing microbial protein to achieve their major target of yielding or making available the end product of digestion i.e. volatile fatty acids (VFA) as energy and protein supply to the host animal.

This end products contribute to the nutrition or nutrients the host animal (ruminants) will benefit and in this microbial fermentation process there are wastages associated like loss of methane, loss of ammonia nitrogen, inefficiencies which limits production performances and release of pollutant that can affect the atmosphere [8, 13–15].

Different kinds of studies have been conducted by many researchers to show that plant secondary metabolites (PSM) manipulate rumen ecosystem by inhibiting, suppressing and proliferation of some microbes (i.e. gram-positive or gram-negative bacteria) and at times help in defaunating (i.e. removal or reduction in the population of protozoa) in the rumen [1]. This aspect of rumen manipulation for proper feed efficiency/utilization and mitigation of methane gas from the host animal has now generated a lot of interest.

2.2 Rumen manipulation

Rumen manipulation is the modification of rumen fermentation processes, so as to improve protein and carbohydrate metabolism and at the same time reduce ammonia, carbon dioxide, hydrogen, methane production and release to the atmosphere. Rumen manipulation aims at minimizing the role of rumen microbes in fermenting ingested feed thus improving the efficiency of nutrient utilization, feed energy and nitrogen loss. The manipulation of rumen involves mitigating the fermentative action of rumen microbes on ingested feed thus increasing feed available to the true stomach (abomasum) where the digestive enzymes act on them and are later absorbed in the small intestine. Many researchers have been carried out by various researchers on the manipulation of rumen microbial ecosystem in order to improve the productivity of ruminants [7, 8, 16].

The increase in the nutrient requirement of ruminants determines changes in the composition of a feed ration. In order to maintain high production level and for the synthesis of microbial protein, ruminants require reduced proportion of green forage rich in structural carbohydrates (such as cellulose), increased proportion of starch and increased proportion of nitrogen and exogenous amino acids in a feed ration [17].

The commensalistic relationship between the host animal (polygastric) and microorganisms accommodated in the rumen becomes disturbed for a grazing and nomadic animal. The high request for nutrients, mainly carbohydrates and protein results in low symbiotic effectiveness between the animals and the microorganisms; despite the increased demand, animals are still not able to utilize excessive amounts of protein and energy. The non-degraded protein and protein digestion leads to the increased rumen ammonia (NH_3) production, while the disturbed proportions of feed (carbohydrates and changes in their fermentation) result in the increased methane (CH_4) production.

Both the aforementioned gases (NH_3 and CH_4) belong to the group of gases called greenhouse gases (GHG) whose level of production in the rumen and the amount emitted to the atmosphere is linearly dependent on the composition of a feed ration [17]. Rumen manipulation also helps in keeping a low hydrogen pressure in the rumen by reducing carbon dioxide thus reducing methane production. Protozoa play a negative role in protein availability and utilization by ruminants [18] by consuming and digesting a substantial number of ruminal bacteria thus reducing the amount of bacterial protein available for enzymatic digestion in the duodenum [19, 20].

Protozoans have the ability to perform the processes of proteolysis and deamination and tampering with the rumen to eliminate the protozoan populations in the rumen which is been referred to as defaunation, which may results in an increase in the amount of nitrogen (microbial source).

Thus, it can be said that the main objective of rumen manipulation is to improve rumen fermentation processes, improve feed efficiency and utilization, reduce nitrogenous wastage, and reduce methane production with emission into the environment thereby ensuring that the total energy available to ruminants is not reduced. Methane (CH_4) production/formation is a product of an enteric fermentation in the rumen of ruminant animal which is widely referred to as a loss of feed energy and suppressing its formation is a very big challenge to ruminant scientist (nutritionist and rumen manipulators).

Ruminants have evolved over thousands of decades to utilize cellulose and polysaccharides by means of a (foregut) pre-gastric fermentation system which yields methane and there is no system to halt methane production [21]. Thus, production of methane (feed energy) in the rumen and its release into the atmosphere, decreases feed utilization. The energy loss derived from the process of feed fermentation is reported to be between 2 to 12% of feed gross energy [22]. Patra *et al.* [23] also reported that methane loss represents about 12% of the gross energy of feed fed to the animals. Donald and Ward [24] also reported that about 95% of the global animal enteric methane is from ruminants which are a consequence of their large population, body size and feed intake. Hence, decreasing the production of enteric methane in ruminants without altering productivity in the animal is desirable both as a strategy to reduce global greenhouse gases emissions and as a means of improving feed conversion efficiency [25, 26].

The process of methanogenesis occurs mainly in the rumen in the presence of microorganism and in an anaerobic environment. The (host) animal provides necessary nutrients needed and environmental conditions that is suitable for the fermentative activities in degrading the solids or substrates (carbohydrate and protein) by the microorganisms thereby synthesizing microbial protein in order to supply energy and protein to the host animal [7, 27].

As a result of interspecific symbiosis between methanogens and bacteria, protozoa or fungi, transfer of hydrogen originating from the cells of the above-mentioned microorganisms to methanogens occurs. Methanogens use CO_2 in the reduction of H_2 and the energy obtained in the process is used for the formation of Adenosine triphosphate (ATP). Owing to the process of methane production in the rumen, low concentration of H_2 is maintained in the rumen environments and this probably affects carbohydrate transformations in the rumen [28, 29].

The methane produced must be expelled together with CO_2 through the process of eructation. Methanogens identified to be present in the rumen of ruminant are of the genus *Methanobrevibacter* and *Methanosarcina.* The main species are: *Methanobrevibacter ruminantium, Methanosarcina barkeri, Methanosarcina mazei and* [30]. However, the development of modern molecular techniques establishes that a ruminant species determines the type of methanogen that prevails in their rumen.

Many authors have suggested the implementation of strategies that can help to mitigate the adverse effects of the process of methanogenesis which occurs in the rumen of ruminant animals on the environment. These mitigation effects can be achieved directly by decreasing the amount of emitted methane per unit of consumed feed and indirectly by increasing the animal performance with the same level of methane emission [31]. Similarly, the reduction of hydrogen production can be achieved without any adverse effect on feed digestion. Alternatively, more favorable utilization of hydrogen by host (animal) and reduction of both the number and the activity of methanogens [26] could go a long way to mitigate the adverse effects of methanogenesis.

Many research centres in the world conduct research on the methods of reducing methane emission to the atmosphere. Lately the interest in the use of phytofactors to achieve the aim has increased. While the current focus is specifically on diminishing methane production from digested feed. It should be noted that from an environmental viewpoint, the final interest lies in controlling methane emission of the entire system and this entails several contexts, some of which are well suited to human intervention [7, 32, 33]. For example, feeding practices which increase feed efficiency of ruminants will ultimately decrease emissions of methane per unit animal product [34]. Faniyi *et al.* [8, 33] and Monteny *et al.* [35] explained that the rate of methane production by ruminants depends on the level of feed intake and the fraction of ingested energy lost, as methane is being reduced with higher feed intake. This is mainly due to increased passage rate in the rumen. The focus now is on nutritional intervention aimed specifically at controlling the yield of methane per unit feed ingested which is referred to as the relative yield of methane. Despite the rigidity of the rumen towards suppressing methanogenesis, it is possible to reduce the yield of methane. There are two main complementary approaches to effectively reducing methane production:

The first approach takes advantage of the reliability that not all feed components ferment in the same way in the rumen thereby yielding different quantities of methane per unit carbohydrate fermented. It is often assumed that concentrates yield relatively less methane than forages per MJ (mega joules) of GE (gross energy) intake [22, 36, 37]. It is well known that the rapid degradation of carbohydrates leads to the production of volatile fatty acids (VFAs). In the synthesis of this VFAs, hydrogen is produced, much of which undergo chemical reactions with methanogenic bacteria thereby leading to methane production. The formation of volatile fatty acids among feedstuffs and diets determines the amount of excess hydrogen in the rumen, which is ultimately converted to methane by methanogenic bacteria. Thus, replacing structural fiber with non-structural carbohydrates shifts volatile fatty acid formation patterns to less of acetic acid and more of propionic acid formation. Therefore, increase in dietary starch at the expense of fiber in a ruminant ration reduces the loss of methane per MJ of GE intake by redirecting or reducing equivalents from methane to propionate production [38]. Related to this approach is the administration of some feed additives such as dietary enzymes or probiotics, which potentially enhance digestion and consequently reduce rumen methane production. It has been demonstrated that the addition of enzymes such as Cellulases and hemicellulases to the diet of ruminants have reduced *in vivo* methane production by 9 and 28% respectively possibly by reducing the acetate to propionate ratio [39].

A second approach involves the use of specific ingredients or additives aimed at specifically reducing production of methane. These are compounds which directly or indirectly inhibit methanogen function. Several chemicals inhibit methane production experimentally (e.g. several halogenated methane analogues such as chloroform and bromochloromethane, [40], and statins [41]. However, these substances have drawbacks as many cause only a transient decline in methane production and they are toxic to the host. Some plant secondary metabolites and plant extracts fall in this category e.g. anthraquinones, which is a major secondary compound of rhubarb, depress rumen methane production [8, 27, 32, 36, 42, 43]. Fatty acids particularly medium chain fatty acids such as myristic and lauric acids preferentially inhibit methanogenic bacteria [39]. Host's immunization against its methanogens has also been examined [44] with favorable but unrepeated results.

Although, there are substances such as ionophores and some plant extracts which cause indirect inhibition of methane formation by causing impairment in the microorganism's habitat or availability of substrates to methanogens (e.g., tannins and saponins). This approach is less dependent on providing alternative hydrogen sinks. Ionophores decrease hydrogen availability and therefore methane production [7, 12, 36, 45]. Some plant extracts such as tannins, saponins and so on also cause differential inhibition of some bacterial species acting in a similar way as ionophores [13].

3. Fermentation in soil systems

3.1 The soil system

The soil systems comprise of organic (5%), inorganic (45%), water (25%) and air (25%) components. These components are always in equilibrium. However, the soil serves as repository for all forms of waste, be it solid, organic and inorganic. The organic matter content of the soil is the portion that contains plant tissues and animal remains occurring at different decomposition stages. Majority of agriculturally productive soils contain about 5% content of organic matter.

Organic matter comprises of components in three major categories:

i. Plant residues and living microbial biomass.

ii. Detritus (often referred to as active soil organic matter).

iii. Humus, which is the stable component.

The first category comprises of microorganisms which break down residues of plants and animal remains as well as detritus. Humus represents the stable portion that is resistant to further degradation, hence, it is the final product of decomposition.

The first two categories of organic matter play a significant role in determining the fertility status of soil. This is because their breakdown account for the mineralized nutrients (such as NO_3^-, PO_4^{2-}, Ca^{2+}, Mg^{2+} etc.) available for the nutrition of crop-plants. The humus component has little contribution to the fertility status of soil, and it is called "stable organic matter". However, it is still very relevant to soil fertility management because it enhances the structure and tilth of soil as well as providing surfaces for cation exchange. Humus is responsible for the soil's dark coloration. The most important way of adding organic matter to the soil system is by the process called composting.

3.2 Composting

The term "composting" is used worldwide with differing meanings. It was defined narrowly by some textbooks as aerobic form of decomposition mainly by aerobic or facultative microbes.

There are two categories of composting, consequent upon the mode of decomposition.

These categories include anaerobic composting, often referred to as *dry fermentation* or anaerobic digestion [46] and aerobic composting, often referred to as *aerobic fermentation* [47].

In anaerobic composting (also known as anaerobic digestion), decomposition occurs where oxygen is absent or in limited supply. During this process, anaerobic micro-organisms (mostly bacteria) play prominent roles in the breakdown of substrates resulting in production of intermediate by-products such as organic acids, methane, and other gases. As the bacteria "work," they generate biogas. Generally, different materials exhibit different digestibility and the more digestible the organic matter is, the more biogas is produced. In the absence of oxygen, the intermediate compounds accumulate and are not further metabolized. These (intermediate by-products that are not fully oxidized) present some phytotoxic properties and very pungent odors. Anaerobic composting occurs under a low-temperature condition; hence, the process does not eliminate pathogens and seeds of weeds. The process also occurs at a slower pace than aerobic composting. These drawbacks offset the advantages of the process, such as: low energy requirement and no loss of nutrients during the process.

Aerobic fermentation or composting, on the other hand, occurs where oxygen is available. The facultative or aerobic bacteria are involved in the breakdown of substrates to release some plant nutrients, heat, biogases, and stable materials (humus). Although intermediate by-products are also produced during aerobic composting, these by-products are further oxidized to yield some useful ions or nutrients for plant growth with little or no danger of phytotoxicity, free of odor and leaving materials resistant to decomposition such as lignified materials (cellulose and hemi-cellulose). The resultant end-product is regarded as compost. The heat generated (due to high temperature regime) facilitates decomposition process within a very short time frame. Moreover, this process destroys many micro-organisms that can cause diseases to crop-plants and humans, as well as weed seeds, owing to the sufficiently high temperature. There are tendencies for loss of nutrients under this process, but it remains an efficient and a more useful method of composting than the anaerobic process for agricultural production and productivity.

Composting of waste is a form of *aerobic fermentation* mode of decomposing solid wastes under controlled conditions of pH, moisture contents, particle size, C/N ratios, etc. [47]. The process resulted in the formation of humus, usually regarded as compost, which serves as a source of nutrients to crop-plants. It involves the accumulation of organic waste in a form of heap. Usually, the waste materials can be shredded to manage the particle size. The accumulation is a multi-layered heap of organic wastes in a windrow, subjected to regular turning to ensure good aeration; and addition of water in order to regulate temperature (heat). Facultative bacteria and fungi feed on the substrate to ensure decomposition and release of ammonium ions and other nutrients required by crop-plants for growth and development.

4. Conclusion

Fermentation is a beneficial process in the field or study of agriculture. It is of great importance in animal and plant nutrition systems. In animal nutrition, the process is basically anaerobic and requires a process called rumen manipulation to enhance better productivity of ruminant animals. However, fermentative processes can either occur aerobically or anaerobically in soil systems resulting in the release of nutrients for the enhanced productivity of crop-plants. Hence, fermentative process involves the transformation of raw materials (organic or inorganic substrates), whether aerobically or anaerobically, to other valuable products through the activities of microorganisms. This implies that fermentation could occur either in aerobic (presence of oxygen) or anaerobic (absence of oxygen) conditions.

Author details

Tolulope Oreoluwa Faniyi and Olukayode Stephen Oyatokun*
Department of Crop and Animal Science, Faculty of Agriculture, Ajayi Crowther, Oyo, Oyo State, Nigeria

*Address all correspondence to: os.oyatokun@acu.edu.ng

References

[1] Santra, A., Saikia, A. and Baruah, K. K. 2012. Scope of rumen manipulation using medicinal plants to mitigate methane production. *Journal of Pharmacognosy,* 3(2): 115-120.

[2] Wanapat, M. 2000. Rumen manipulation to increase the efficient use of local feed resources and productivity of ruminants in the tropics. *Asian-Austr. J. Anim. Sci.,* 13: 59-67.

[3] Frederique, Chaucheyras-Durand, Eric, Chevaux, Cecile, Martin and Evelyne, Forano (2012) Use of Yeast Probiotics in Ruminants: Effects and Mechanism of Action on Rumen pH, Fibre Degradation, and Microbiota According to the Diet. *Intechopen* DOI: 10.5772/50192

[4] Patra, A. K. and Saxena, J. 2009. A review of the effect and mode of action of saponins on microbial population and fermentation in the rumen and ruminant production. *Nutr. Res. Rev.*, 22: 204-219.

[5] Babayemi, O. J. (2007). *In vitro* fermentation characteristics and acceptability by West African dwarf goats of some dry season forages. *African Journal of Biotechnology*, 6(10): 1260-1265.

[6] Vilela, A. (2019) The Importance of Yeasts on Fermentation Quality and Human Health-Promoting Compounds :Review *Fermentation* 5,(46); doi:10.3390/fermentation5020046

[7] Faniyi, T. O., Adegbeye, M.J., Elghandour,M.M.M. Y., Pilego, A. B., Salem, A. Z. M., Olaniyi, T. A., Adediran, O. and Adewumi, M. K.(2019) Role of diverse fermentative factors towards microbial community shift in ruminants. *Journal of Applied Microbiology* . https://doi:10.1111/jam.14212 ISSN 1364-5072

[8] Faniyi, T. O., Adewumi, M. K. Prates, E. R. and Ayangbenro, A. S. (2016) Effect of herbs and spices (Plant Extracts) on rumen microbial activities: A Review. Publicações em Medicina Veterinária e Zootecnia.v.10, n.6, p.465-474

[9] Kongnum, P., Wanapat, M., Pakdee, P. and Navanukraw, C. (2009). Effect of coconut oil and garlic powder on *in vitro* fermentation using gas production technique. *J. Livestock Science,* 127(1): 38-44.

[10] Manh, S. N., Wanapat, M., Uriyapongson, S., Khejornsart, P. and Chanthakhoun, V. 2012. Effects of eucalyptus (*Camaldulensis*) leaf meal powder on rumen fermentation characteristics in cattle fed on rice straw. *African Journal of Agricultural Research*, 7(13): 1997-2003.

[11] McGuffey, R. K., Richardson, L. F. and Wilkinson, J. I. D. 2001. Ionophore for dairy cattle: Current status and future outlook. *J. of Dairy Science,* 84: 194-203.

[12] Russell, J. B. and Houlihan, A. J. 2003. Ionophore resistance of ruminal bacteria and its potential impact on human health. *FEMS Microbiol. Rev.*, 27: 65-74.

[13] Calsamiglia, S., Busquet, M., Cardozo, P. W., Castillejos, L. and Ferret, A. (2007). Essential oils as modifiers of rumen microbial fermentation. *Journal of Dairy Science,* 90(6): 2580-2595.

[14] Hook, S. E., Wright, A. D. G. and McBride, B. W. (2010). Methanogens: methane producers of the rumen and mitigation strategies. Archaea, 11, http://dx.doi.org/10.1155/2010/945785 (article ID 945785).

[15] Mohammad, E. N. S. and Mohammad, M. M. 2015. The Influence

of Ginger (*Zingiber officinale*) on *in vitro* Rumen Fermentation Patterns. *Annual Research and Review in Biology*, 5(1): 54-63.

[16] Taghizadeh, A., Alizadeh, S. and Noubakht, A. 2010. Effect of lasalocid on ruminal and blood metabolities and performance of gizel male lambs. *Archive of SID*, 68-78.

[17] Szumacher-Strabel, M. and Cieslak, A. 2010. Potential of phytofactors to mitigate rumen ammonia and methane production. *Journal of Animal and Feed Sciences,* 19: 319-337.

[18] Benchaar, C., Calsamiglia, S., Chaves, A. V. Fraser, G. R., Colombatto, D., McAllister, T. A. and Beauchemin, K. A. (2008). A review of plant-derived essential oils in ruminant nutrition and production. *Anim. Feed. Sci. Technol.*, 145: 209-228.

[19] Ivan, M., Dayrell, M. D., Mahadevan, S. and Hidiroglou, M. (1992). Effect of bentonite on wool growth and nitrogen metabolism in fauna-free and faunated sheep. *J. Anim. Sci.,* 7: 3194-3202.

[20] Ivan, M., Neill, L., Forster, R., Alimon, R., Rode, L. M. and Entz, T. (2000). Effects of Isotricha, Dasytricha, Entodinium and total fauna on ruminal fermentation and duodenal flow in wethers fed different diets. *J. Dairy Sci.*, 83: 776-787.

[21] Gill, M., Smith, P. and Wilkinson, J. M. (2010). Mitigating climate change: the role of domestic livestock. *Animal*, 4: 323-333.

[22] Johnson, K. A. and Johnson, D. E. (1995). Methane emissions from cattle. *J. Anim. Sci.*, 73: 2483-2492.

[23] Patra, A., Park, T. and Kim, M. 2017. Rumen methanogens and mitigation of methane emission by anti-methanogenic compounds and substances. *J Animal Sci Biotechnol* **8,** 13 https://doi.org/10.1186/s40104-017-0145-9

[24] Donald, E. J. and Ward, G. M. (1996). Estimates of animal methane emissions. *Environmental monitoring and assessment,* 42(1-2): 133-141.

[25] Dong, G. Z., Wang, X. J., Liu, Z. B. and Wang, F. (2010). Effects of phytogenic products on *invitro* rumen fermentation and methane emission in goats. *Journal of Animal and Feed Sciences,* 19: 218-229.

[26] Martin, C., Morgavi, D. P. and Doreau, M. 2010. Methane mitigation in ruminants: from microbe to the farm scale. *Animal*, 4: 351-365.

[27] Akaninyene Asuquo Jack, Michael Kolawole Adewumi, Okpara Oghenesuvwe, Moyosore Joseph Adegbeye , Daniel Ekong Ekanem, Tolulope Oreoluwa Faniyi (2020) Effect of water-washed neem (*Azadirachta indica*) fruit on rumen digesta fatty acids and biohydrogenation intermediates of fattened West African dwarf rams International *Journal of Environment, Agriculture and Biotechnology*, 5(5),Pg.: 1329-1337, ISSN: 2456-1878 https://dx.doi.org/10.22161/ijeab.55.20 Sep-Oct, 2020 | Available: https://ijeab.com/

[28] Ushida, K. and Jouany, J. P. 1996. Methane production associated with rumen ciliated protozoa and its effect on protozoan activity. *Letters Appl. Microbiol.,* 23: 129-132.

[29] Wolin, M. J. and Miller, T. L. 1988. Microbe-microbe interaction. In: P.N. Hobson (Editor). The Rumen, Microbial Ecosystem. *Elsevier Applied Science,* London. pp. 343-459.

[30] Stewart, C. S., Flint, H. J. and Bryant, M. P. 1997. The rumen bacteria. In: The Rumen Microbial Ecosystem

(Ed. P. N. Hobson and C. S. Stewart). pp. 10-72. Blackie Academic and Professional Publishers, London.

[31] Kreuzer, M. and Soliva C. R. (2008). Nutrition: key to methane mitigation in ruminants. *Proc. Soc. Nutr. Physiol.*, 17: 168-175.

[32] Adegbeye, Moyosore J.,Elghandour, Mona M. M. Y., Faniyi, Tolulope O., Perez, Nallely Rivero, Pilego, AlbertoBarbabosa, Zaragoza-Bastida, Adrian and Salem, Abdelfattah Z. M. (2018) Antimicrobial and antihelminthic impacts of black cumin, pawpaw and mustard seeds in livestock production and health. Agroforestry System. Springer Nature B.V. https://doi.org/10.1007/s10457-018-0337-0

[33] Faniyi, T. O., Prates, Ê. R., Adegbeye, M. J., Adewumi, M. K., Elghandour, Mona M. M. Y. , Salem, A. Z. M. Luciano A. Ritt, Angel Sánchez Zubieta, Laion Stella, Elvis Ticiani and Akaninyene A. Jack (2019) Prediction of biogas and pressure from rumen fermentation using plant extracts to enhance biodigestibility and mitigate biogases. Enviromental Science and Pollution Research 1 – 9,https://doi10.1007/s11356-019-05585-1

[34] Capper, J. L., Cady, R. A. and Bauman, D. E. (2009). The environmental impact of dairy production: 1944 compared to 2007. J. Anim. Sci. 87: 2160-2167.

[35] Monteny, G. J., Bannink, A. and Chadwick, D. 2006. Greenhouse gas abatement strategies for animal husbandry. *Agr. Ecosyst. Environ.* 12: 163-170.

[36] Faniyi, T. O., Prates, E. R., Adewumi, M. K. and Bankole, T. (2016) Assessment of herbs and spices extracts/meal on rumen fermentation: Review. Publicações em Medicina Veterinária e Zootecnia.v.10, n.5, p.427-438

[37] Ferris, C. P., Gordon, F. J., Patterson, D. C., Porter, M. G. and Yan, T. (1999). The effect of genetic merit and concentrate proportion in the diet on nutrient utilization by lactating dairy cows. *J. Agric. Sci.,* 132: 483-490.

[38] Benchaar, C., Pomar, C. and Chiquette, J. (2001). Evaluation of dietary strategies to reduce methane production in ruminants: a modelling approach. *Can. J. Anim.Sci.*, 81: 563-574.

[39] Beauchemin, K. A., Kreuzer, M., O'Mara, F. and McAllister, T. A. (2008). Nutritional management for enteric methane abatement: a review. *Aust. J. Exp. Agric.*, 48: 21-27.

[40] McAllister, T. A. and Newbold, C. J. 2008. Redirecting rumen fermentation to reduce methanogenesis. *Aust. J. Exp. Agric.*, 48: 7-13.

[41] Miller, T. L. and Wolin, M. J. 2001. Inhibition of growth of methane-producing bacteria of the ruminant forestomach by hydroxymethylglutaryl-SCoA reductase inhibitors. *J. Dairy Sci.*, 84: 1445-1448.

[42] Akaninyene A. Jack, Michael K. Adewumi, Moyosore J. Adegbeye, Daniel E. Ekanem, Abdelfattah Z. M. Salem and Tolulope O. Faniyi 2020. Growth-promoting effect of water-washed neem (Azadirachta indica A. Juss) fruit inclusion in West African dwarf rams. *Tropical Animal Health and Production* https://doi.org/10.1007/s11250-020-02380-w

[43] García-González, R., González, J. S. and López, S. (2010). Decrease of ruminal methane production in Rusitec fermenters through the addition of plant material from rhubarb (*Rheum spp.*) and alder buckthorn (*Frangula alnus*). *J. Dairy Sci.*, 93: 3755-3763.

[44] Wright, A. D. G., Kennedy, P., O'Neill, C. J., Toovey, A. F., Popovski, S.,

Rea, S. M., Pimm, C. L. and Klein, L. 2004. Reducing methane emission in sheep by immunization against rumen methanogens. *Vaccine,* 22: 3976-3985.

[45] Moyosore Joseph Adegbeye, Mona M.M.Y. Elghandour , Jose Cedillo Monroy, Taye Olurotimi Abegunde, Abdelfattah Z.M. Salem, Alberto Barbabosa-Pliego, Tolulope O. Faniyi (2019) Potential influence of Yucca extract as feed additive on greenhouse gases emission for a cleaner livestock and aquaculture farming - A review / *Journal of Cleaner Production* 239 https://doi.org/10.1016/j.jclepro.2019.118074

[46] Carstensen, F.V.,Graziano, M.,Vorotytseva, N.Waite, W.E. and Parr. K.E. 2013. Economic Analysis of Bioenergy: An Integrated Multidisciplinary Approach. In New and Future Developments in Catalysis. Pp 297 -323.doi.org/10.1016/8978-0-444-53878-9-00015-1

[47] El-Haggar, S.M. 2007. Sustainability of Agriculture and Rural Wastes in: Sustainable Industrial Design and Waste Management: Cradle-to-Cradle for Sustainable Development. Elsevier Academic Press, Cambridge, MA. Pp 223-260

Chapter 2

Strategies for Enhancing Product Yield: Design of Experiments (DOE) for *Escherichia coli* Cultivation

Puneet Kumar Gupta and Jyotheeswara Reddy Edula

Abstract

E. coli is considered one of the best model organism for biopharmaceutical production by fermentation. Its utility in process development is employed to develop various vaccines, metabolites, biofuels, antibiotics and synthetic molecules in large amounts based on the amount of yield in shake flasks, bioreactors utilised by batch, fed-batch and continuous mode. Production of the desired molecule is facilitated in the bioreactor by employing strategies to increase biomass and optimised yield. The fermentation is a controlled process utilising media buffers, micronutrients and macronutrients, which is not available in a shake flask. To maximise the production temperature, dissolved oxygen (aerobic), dissolved nitrogen (anaerobic), inducer concentration, feed or supplementation of nutrients is the key to achieving exponential growth rate and biomass. Design of experiments (DOE) is critical for attaining maximum gain, in cost-effective manner. DOE comprises of several strategies likewise Plakett-Burman., Box–Behnken, Artificial Neural Network, combination of these strategies leads to reduction of cost of production by 2–8 times depending on molecules to be produced. Further minimising downstream process for quickly isolation, purification and enrichment of the final product.

Keywords: *E. coli*, Design of experiments (DOE), Response surface methodology, Bacterial Fermentation, Batch, Fed-batch, Process optimisation, glycosylation, disulfide bridges, codon optimisation, host, post-translational modifications, inclusion bodies

1. Introduction

E. coli is most studied bacteria learned of symbiotic relationships with human for years derived after culturing F Plasmid of a 1922 isolate from a diptheria patient. Production of biopharmaceuticals from *E. coli* is in practice since 1965. Also, the it served for production of biopharmaceuticals such as recombinant proteins, metabolites by several companies namely BPB Bioscience, Agilent technologies, Promega, Takara, Tonbo Biosciences, New England Biolabs, Novagen and Lucigen. They are optimised for expression of challenging proteins difficult to express, purify and folding in native conformation. Moreover, formation of proper disulphide bonds and refolding of membrane proteins is also achieved by using newly commercially available strains. These strains are suitable for production of metabolites and enzymes in easy to scale-up process development. The cultivation of *E. coli* is facilitated by optimisation of concentrations of tryptone (plant and animal derived), yeast extract, casamino acids. These are divided in three categories carbon, nitrogen and micronutrients, as per sustainable development goals set by united nations for 2030. Innovation for increasing productivity using technology is one of the 17 subject areas, it is need to follow typical strategy, which is usually unique for production of an active compound or protein. The **Table 1**, below describes different strategies to discovered for optimisation of product by utilising model organism *E. coli* (**Table 1**) [57].

Production of Biopharmaceuticals and Biomolecules during fermentation require media formulated using cheap ingredients for supplementation of carbon and nitrogen. The carbon sources are typically glucose, glycerol, molasses (**Table 1**). Nitrogen sources are typically yeast extract and digested proteins by enzymes. The utilisation of carbon by *E. coli* during batch can be inhibited due to accumulation of acetate, formate and succinate. Historically several attempts have been made to design defined or semi-defined media recipes that do not require bio nutrient additives. Traditionally, media developed generally utilise one-factor-at-a-time (OFAT) where all but one ingredient (factor) is fixed [58, 59]. This approach guides towards determining individual factors that influence culture growth. Effect of utilisation of amino acids, vitamins and minerals interaction are not possible using OFAT. These interactions can be studied using DoE and multifactorial designs.

DoE is a multifactorial methodology utilising statistical approaches to design and analyse an experimental process by which better outcome and results are obtained. Bacterial growth is the system response impact of "DoE planning" by varying critical parameters and simultaneous factors, such as phosphate, sulphate, vitamin and amino acid concentrations. The full factorial design, DoE is attributed by important principals to allow for robust, valid experimentation. These are known as treatments, replicated, randomised to protect against hidden factors [59]. All treatments must be performed in blocks to control sources of variation. In an experimental process design, all four principles, the statistical approach, conform to regulations and complete exploration of an extensive experimental design process, helps in reducing the number of experiments to get a good outcome [60]. For cultivation using *E. coli*, statistically relevant experimental designs for industrially important proteins (rB-glucosidase, human superoxide dismutase) is studied. A similar approach is applied enzymes, recombinant antibodies and therapeutic proteins(Ranibizumab, Somatropin Certolizumab pegol, PEG interferon alfa-2b, Romiplostim, Interferon-beta 1b, Pegloticase, etc.) [61]. The optimisation of media and critical components is needed for *E. coli* fermentation in batch and fed-batch fermentation is required to achieve a better growth rate, target product per litre of fermentation and reduced cost of production. The cost of various fermentation ingredients for one batch size 10 L of fermentation ranges from 20 to 155 USD (**Table 2**). The cost of media ingredients reduces after scaling up of process for production from 15 to 5000 L.

The DoE with a correct statistical model to understand the effect is needed to be adopted by industrial producers to achieve a robust and high yielding process. The DOE is statistically planned trials for optimising factors associated with increased product output, yield and cost reduction. Typically, DOE experiments are done to screen up to 2–15 elements. Traditionally, using OFAT, the number of experiments needed to determine the effect on yield, the output is numerous. Using the DOE approach, several components, additives and sources of nutrients are compared

Name of Product/ Molecule	Type of Fermentation	Design optimization	Media	Carbon Source	Reference
1,3-propanediol	Batch	Glycerol and glucose Co-fermentation	Reinforced Clostridium Medium (RCM)	Glucose	[1]
D-amino acid Oxidase	Batch	Combination fermentation	Complex medium	Glycerin	[2]
O-acetylhomoserine	Batch	Glycerol-Oxidative pathway	MS medium	Glycerol	[3]
Heparosan	High cell density	High cell density fermentation	Chemically defined fermentation media	Glucose	[4]
O-succinyl-l-homoserine	Batch	Multilevel fermentation optimization	Defined media	Glucose	[5]
(2S)-Naringenin	Batch	Fermentation and metabolic pathway optimization	Optimised MOPS minimal medium	Glucose	[6]
L-tryptophan	Batch	Batch Fermentation	Complex medium with additives	Glucose	[7]
Beta-farnesene	Batch	Batch Fermentation	2 × YT medium	Glycerol	[8]
L-tryptophan	Batch	Enzymatic hydrolyzate	Complex medium with yeast extract	Glucose	[9]
VP2 VLP vaccine	High cell density	Fed batch	Luria Bertani Media	Glycerol	[10]
Vitamin B12	Batch Fermentation	Metabolic engineering	LB media	Glucose	[11]
L-Valine	Batch fermentation	Two stage fermentation	Complex Medium	Glucose	[12]
PhoA	Batch Fermentation	Design of experiments for *Escherichia coli* fermentation	Complex medium Phosphate optimization	Glycerol	[13]
L-methionine	Batch fermentation	Design of experiments with Genetic engineered Escherichia coli	Complex medium with yeast extract	Glucose	[14]
Reteplase	Batch fermentation	Optimization using response surface methodology	LB medium (Response surface methodology for pH, Temp, RPM)	—	[15]
Tauroursodeoxycholic acid	Batch fermentation	Optimization of engineered Escherichia coli cell factory	Complex medium	Glucose and Glycerol	[16]

Name of Product/ Molecule	Type of Fermentation	Design optimization	Media	Carbon Source	Reference
Tyrosine phenol lyase	Batch fermentation	Statistical medium optimization	Complex medium	Sucrose	[17]
Phytochromes Production	High-cell density *E. coli* fermentation	High cell density fermentation	Minimal Medium	Glucose	[18]
Keratinolytic protease	Batch fermentation	Protein engineering to enhance keratinolytic protease activity	Complex Medium	Glucose	[19]
L-Threonine	Batch fermentation	Biofilm-Based fermentation strategies	Complex Medium	Glucose	[20]
Therapeutic DNA vaccine	Batch fermentation	Response surface methodology	2 x YT medium	Mannitol	[21]
Pyruvate Oxidase	High cell density	Temperature gradient based methodology	Yeast, Tryptone complex medium	Glycerol	[22]
rPDT fusion protein	Piolet scale batch fermentation	Statistical experimental design	LB medium	—	[23]
Platelet aggregation inhibitor	High cell density fermentation	Optimization of production	Yeast, Typtone	Glycerol	[24]
Vitreoscilla haemoglobin	Batch fermentation	Immobilised *E. coli*	Cheese Whey, Yeast extract	—	[25]
D-Lactose	Fed-batch fermentation	Optimization of D-lactose production	Modified mineral salt medium	Glucose	[26]
Curcumin	Batch Fermentation	Optimization of fermentation	TB medium (Tryptone, Yeast extract)	Glycerol	[27]
Alkaline beta-mannanase	Fed-batch fermentation	Induction conditions optimization	TB medium (Tryptone, Yeast extract) with additives	Glycerol	[28]
HIV-1 P17	Fed batch fermentation	Multiparameter optimization for secreted protein	TB medium (Tryptone, Yeast extract)	Glucose	[29]
Heptatitis Delta Virus (HDV) antigen	Fed batch fermentation	Optimization of Fed batch	LB medium	Glycerol	[30]
Chitosanase	Fed batch fermentation	Optimization of secretion efficiency PelB-CSN-N	Terrific Broth with trace elements	Glycerol	[31]
Nitrilase	High cell density	Optimization of temperature, substrate and IPTG	Super Optimal	Glycerol	[32]

Name of Product/ Molecule	Type of Fermentation	Design optimization	Media	Carbon Source	Reference
			broth (SOB) medium		
Human Interferon Beta	Batch fermentation	Reduction of acetate by Response surface methodology	Terrific Broth (TB)	Glucose	[33]
Beta-chitin	Batch fermentation	Selection of strain and IPTG optimization	Luria Bertani (LB)	—	[34]
Beta-glucosidase	Batch fermentation	Ag43-mediated thermostable for high temperature fermentation	Mineral Media	Glucose	[35]
46 KDa Antibody Fab' fragment	Batch fermentation	ultra-scale-down (USD) approach based on focused acoustics	Modified defined medium	Glycerol	[36]
Therapeutic protein production (10 KDa)	Fed-Batch fermentation	Optimization of inclusion bodies by autoinduction without adding IPTG due to galactose in the complex media	Yeast and Soy Peptone	Glycerol, Glucose	[37]
Catechol-O-methyltransferase	Batch fermentation	Optimization of production by artificial neural network	Tryptone, Yeast extract (SOB Medium)	Glycerol	[38]
Platelet aggregation inhibitor, salmosin	Fed-batch fermentation	Optimization of carbon and nitrogen sources	Yeast and Tryptone	Glycerol	[39]
VB4–845, an immunotoxin	High cell density fermentation	Optimization of VB4–845 titre in the supernatant	Glycerol Minimal medium	Glycerol, Arabinose	[40]
GDP-mannose	Coupling fermentation	Optimization by Coupling fermentation	Defined media	Fructose, Mannose	[41]
Heparosan	Fed-Batch fermentation	Defined medium using exponential fed-batch glucose addition with oxygen enrichment	Defined media	Glucose	[42]
Pig liver esterase	Batch fermentation	Optimization of co-expression in combination with chaperones GroEL/ ES allowed the production of soluble and active enzyme	Complex medium	Glucose	[43]
Penta-N-acetyl-chitopentaose	Two step fermentation	Optimization of oligosaccharides production	Minimal Medium	Glycerol	[44]
Resilin	Batch fermentation	Primary induction step before lactose induction	Lactose, Casein, Yeast	Glycerol	[45]

Name of Product/ Molecule	Type of Fermentation	Design optimization	Media	Carbon Source	Reference
Malaria antigen PvRII	High cell density	High cell density for production of recombinant proteins	Defined media	Glucose	[46]
Bacterial phytase	Fed-batch fermentation	Optimization of two different feeding strategies	Terrific Broth	Glucose	[47]
Human interferon-gamma (hIFN-gamma)	High cell density fermentation	Exponential growth by using glucose as carbon source	M9 Modified medium	Glucose	[48]
B subunit of Escherichia coli heat-labile enterotoxin (LTB)	High cell density	Step wise addition of IPTG	Complex media	Glucose	[49]
Plasmid pAQN carrying the aqualysin I (AQI) gene	Batch fermentation	Optimization of yeast extract supplementation	LB and M9 minimal media	Glucose	[50]
Prochymosin	Batch fermentation	Use of lactose as energy source and inducer	Complex medium	Glycerol, Glucose	[51]
HIV-1 protease	Batch fermentation	Separation of growth and production phases in a two-step process	LB and M9CA medium	Glucose	[52]
Porcine somatotropin protein	High cell density	Oxygen enrichment, yeast extract (YE) effect, optimal specific growth in switching on gene expression, and feeding strategies.	Defined media with additives	Glucose	[53]
Human interferon-alpha 1	High cell density	Optimization of nutrient source	Defined media	Glucose	[54]
Human interleukin-1 beta	High cell density	Maintenance of low glucose and acetate for production	Defined media	Glucose	[55]
EcoRI	Batch fermentation	Control of pH and Oxygen	LB media	—	[56]

Table 1.
Strategies and optimization parameters and type of fermentation in production of various proteins and small molecules.

simultaneously, and their impact is evaluated. Based on the response critical factors are ranked. Soon after the response effect by variables are generated, the analysis is successively recorded. Various DoE experiments are studied to understand influencing factors determined by a statistically relevant experimental plan. The difference in factors is measured by Analysis of variance (ANOVA). The results obtained after analysing responses are plotted using contour plots and response surface methodology (RSM). There are several software packages for optimisation of critical factors and parameters Design Expert (Stat-Ease Inc.), GT-SUITE (DOE), DOE ++ (RecCom), MODDE (Umetrics), DoE Fusion PRO (S-Matrix Corp.), STAVEX (Aicos), Minitab (Minitab Inc.), and JMP (SAS) [62].

Component	Batch	Basal and Initial medium	Components added in feed solutions	Cost Per 10 Litre of Fermentation media (USD)
	Batch	Fed batch (Basal)	Fed Batch (Feed Media)	
NH_4Cl	7	7	28	71
KH_2PO_4	1.5	7.5	0	22
Na_2HPO_4	1.5	7.5	0	18
K_2SO_4	0.35	0.85	0	1
$MgSO_4.7\ H_2O$	0.17	0.17	0.86	0
Trace Elements[a]	0.8	0.8	3.2	1
Yeast extract	5	10	15	12
Glucose				0
Batch	1.0–5.0	20		27
Fed-Batch		1	99	1
Glycerol				0
Batch	1.0–3.0			1
Fed-Batch		2	30–60	1
			Total Cost	155[b]

Amount of component added to the medium listed for purpose mentioned (g/L).
[a]Millilitres of trace elements solution, prepared as described added as 1X to medium or feed solution.
[b]Cost of media components calculated with consumption in fermentation using high grade reagents.

Table 2.
Components of media used to grow E. coli in batch and fed-batch fermentations.

2. Critical fermentation ingredients

Critical fermentation ingredients are media components which cannot be replaced. These can only be standardised for maximised yield. Typically, buffers and nitrogen sources such as yeast extract, tryptone are not changed. Since *E. coli* is the most studied and highly utilised systems for producing various enzymes, antibodies, and biological products. Bacteria require specific conditions for growth attributed to factors such as oxygen, pH, temperature, and light. Bacterial growth is divided into lag, exponential (log) and stationary phase. During the initial stage, cellular activity in a rich nutrient medium allows cells to synthesise proteins, cells increase in size, but no cell division occurs in the phase. During the exponential phase, metabolic activity is high as DNA, RNA, cell wall components, or machinery needed for division are generated. The stationary phase is triggered due to the accumulation of waste products and depletion of nutrients. During the late log phase, proteins are induced by the addition of allolactose analogue, Isopropyl β-D-1-thiogalactopyranoside (IPTG) [63, 64]. The expression of recombinant products is controlled by promoter systems like T5 and T7 RNA Polymerases. Alternative promoter systems, such as auto-inducible phoA promoter system [13], the salt-inducible promoter (proU), arabinose-inducible promoter (pBAD) [65], the heat-inducible phage Lamba promoters (pL and pR), the cumate-inducible T5 promoter-based system [66], and the cold-inducible cspA promoter-based system [67] are also valuable for the biologics production. The cost of biologics production is due to the high cost of raw material and fermentation media. In the biologics industry, the more straightforward, cheaper, and reproducible process is highly appreciated.

Fermentation media is a critical component, and a balance of nutrients is needed for increasing productivity. Standardisation of *E. coli* fermentation requires identifying a combination of various media components available, e.g., Yeast extract, Soyabean meal, Bactotryptone, Meat extract and Enzymatic digest of plant and animal protein (Trypic or casein enzymes). There are various carbon sources (glucose, glycerol, sucrose, lactose etc). Additives for fermentation are vitamins, amino acids, and trace elements. Designing a media needs to evaluate the requirement of each of the individual component along with the additives. The design of model using statistical approach having multiple parameters in consideration, followed by validation of defined parameters using fermentation. This is achieved using DoE experimentation (**Figure 1**). Experiments are carried out at Shake flash level with selected nutrients such as carbon and nitrogen sources. Small scale studies are carried out to define as batch or fed-batch fermentation. Next stage is to screen the components available for fermentation of batch/fed batch. Once the components are finalised the possibility of Scale-up is evaluated based on the availability from the source. Finalisation of media components is carried out using shake flask with DOE of media buffer additives and inducers. Evaluation of various product outcome biomass ratio and validation of protein quality is also study with 3 to 5 selected designs. Once the nutrient and components are finalised pilot scale batches setup to study biomass to product ratio. Further optimisation of dissolved oxygen and temperature in fermentation is carried out by the DOE approach. If results are not reproducible with the selected condition, other near possible designs are studied to finalise the medium and process for fermentation (**Figure 1**). The process of selection of components is based on outcome in an experiment calculated by Biomass (OD), product output g L^{-1} and cost of ingredients. The process is clearly defined in (**Figure 1**).

2.1 Batch fermentation

In microbial batch, fermentation cultivation is done in a fixed volume of medium in a fermenter. The standard inoculum in the fermenter is 50–200 mL of shake flask volume in 2–5 L of fermentation media. The batch fermentation

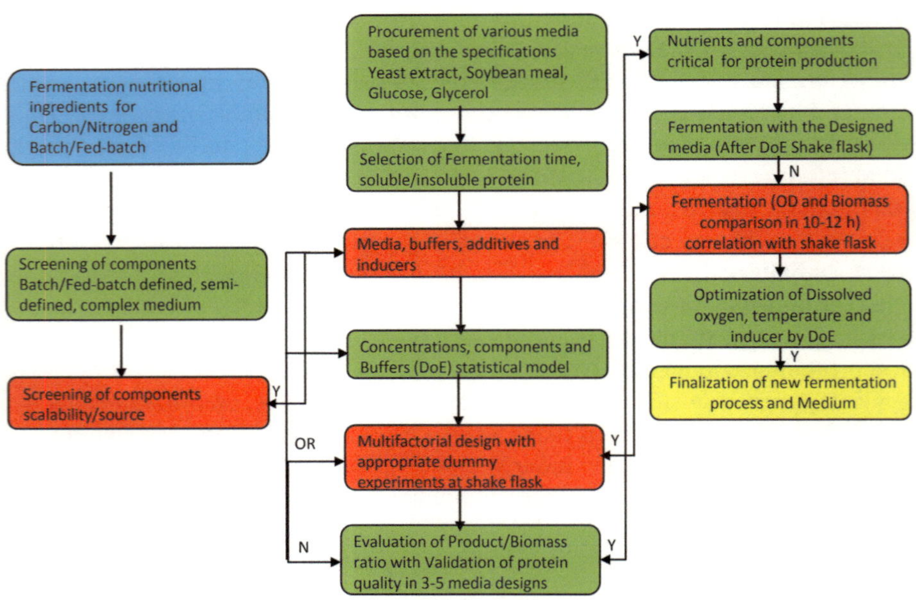

Figure 1.
Flow diagram for logical designing of upstream process for fermentation for E. coli. *Legend: Y = yes, N = No.*

typically OD_{600} is 20–40 in 8–12 h time. The microbial growth depletes the nutrients resulting in the accumulation of by-products; there is a continuous change in the culture environment. After completion of the batch, media and cells are harvested. The advantages are batch fermentation, ease of operation, low risk of contamination, high yield of protein to biomass in less time of fermentation, and majorly for soluble or excreted proteins. Typical disadvantages are relatively long downtime between batches due to vessel setup and sterilisation, low cell/biomass densities, due to cleaning. DOE is needed to optimise the required nutrients and minimise product accumulation during fermentation. Typical batch fermentation media constitutes Yeast extract, Bactotyptone (or Soybean meal) 10 g^{-1} to 24 g^{-1}, respectively. Buffers of Sodium and potassium phosphate in combination to reach pH 6.8. to 7.0, 100 X amino acid solution, Trace elements (400x) is $CoCl_2$. $6H_2O$ (1 mg/ml^{-1}), $MnCl_2.4H_2O$ (6 mg ml^{-1}), $CuCl_2.2H_2O$ (0.6 mg ml^{-1}), H_3BO_3 (1.2 mg ml^{-1}), $Na_2MoO_4.2H_2O$(1 mg ml^{-1}), Zn(II)acetate.$2H_2O$ (5.2 mg ml^{-1}) and Fe(III)citrate (40 mg ml^{-1}) [54]. The typical medium components are listed in the (**Table 2**) for batch fermentation as a base design to start optimisation.

2.2 Fed-batch fermentation

Fed-batch fermentation is a standard mode of fermentation in the bioprocess industry. Typically, fed-batch fermentation starts at the end of batch fermentation. *E. coli* is adapted and cultivated in defined media. In Fed-batch fermentation, cells are inoculated and grown in batch mode for 10–15 h. Once all the nutrients are depleted, evident by analysing the amount of glucose in the medium, dissolved oxygen levels are increased to 60–80%. The Fed-batch fermentation is initiated by starting to feed of Glucose, Vitamins, amino acids, and trace elements. The feed is added to the medium to allow the volumetric cell to increase the mass concentration exponentially. The growth rate of is changed to 0.12–0.22 μ h^{-1} during fed batch stage. These equations determine the growth rate in the medium.

$$ms(i) = (set + mYX/S)VSo, \quad (1)$$

$$ms(t) = ms(i)eset * (t - ti) \quad (2)$$

The first equation, ms (i), is the value of feed rate at the initiation of the fed-batch phase at time ti. μ_{set} is the specific growth rate, m is specific maintenance coefficient, $Y_{X/S}$ is yield coefficient, V is the bioreactor volume, and So is the initial glucose concentration. In the second equation, ms(t) is the rate of addition of substrate (g $^{h-1}$). After induction of protein expression, the specific growth rate of *E. coli* is typically reduced to 0.1 μ $^{h-1}$. The cells are harvested after completion of the run. Suppose the growth rate is not specified during fermentation, constant accumulation of several toxic metabolites produced during the fed-batch process acetate, formate, succinate, and lactate, resulting in oxygen limitation fed-batch. Therefore, it is recommended to wash cells with Tris-EDTA buffers after washing *E. coli* cells are stored or lysed for downstream processing. The distinct advantage of resuspending *E. coli* after completion of batch reduces protein degradation due to metalloproteases [46].

Start of feed is determined by measuring the concentration of substrate in the fermentation broth typically after 10–12 hrs of batch. The feeding strategy should be designed so that the growth rate is maintained to limit the production of toxic formate, acetate and other metabolic compounds, enhancing bacterial growth. The growth of bacteria and conversion of feed to biomass is maximum when the exponential growth phase is maintained. The utility of fed-batch and importance is obtaining high cell density and biomass, leading to increased production of the high amount of product yield. The fed-batch is applicable to increase product yield by limiting growth rate and controlled substrate utilisation (**Table 2**). The media for batch and fed-batch fermentation is listed (**Table 2**).

3. Culture nutrient optimization strategies

3.1 One factor at a time- classical media optimisation methods

Selection of one-factor-at-a-time (OFAT) is a traditional method for optimisation of media. In this strategy only one factor is varied keeping all other parameters constant. The usual choice is ease and convenience; it makes OFAT the most preferred choice for formulating, designing, optimising, and scaling up the fermentation medium [68]. This method is still popular among many research groups for developing the medium for fermentation. Physical parameters, supplementation, removal, replacement and feedback experiments are primary considerations during OFAT. They comprise of growth temperature, operating pressure, size of nutrients or extracts. The bioprocess in fermentation controls constant supply of nutrients, removing metabolic products and toxic compounds, and constantly disseminating the nutrient solids, buffers and salts in liquid and gaseous phases. There is a constant evolution of design improvement for agitation and aeration; these allow better control over flow dynamics, the minimal effect of viscosity, and even circulation of components and nutrients. The healthy growth of culture in the batch is maintained by supplementing nutrients such as M9 medium with $FeSO_4{\bullet}7H_2O$; M63 medium with KOH; A medium $MgSO_4{\bullet}7H_2O$, 20% Glucose or sugar, vitamins, casamino acids or L-amino acids [69]. Removal experiments are required for the identification of critical components needed for the media. Certain media affect the reduction of formate, acetate, and reduction of pH. There are few examples associated with removing glucose from complex media to prevent inhibition of bacterial growth. Replacement experiments identifies correct nutrients complexes for nitrogen source yeast, soy peptone, bactotryptone, meat extract and protein powder. The carbon sources utilised are glucose, glycerol, sucrose, lactose and others. The use of OFAT for designing of media for fermentation limits number of experiments, the approach is suitable for production of metabolite. In one study, precursor carbohydrate phosphotransferase system (PTS) encoding genes a vital DXP pathway were deleted. This resulted in the enrichment of Isoprenoid phosphoenolpyruvate. Growth medium and production of lycopene (a C40 isoprenoid) resulted in maximisation by these culture conditions [70].

Defined media recipes used in the fermentation of *E. coli* include nine mineral salts in usually salts of ammonium, potassium and sodium cations; and carbonate, chloride, nitrate and sulphate anions. Glucose, Glycerol and ammonia were identified as potential additional sources of carbon and nitrogen, respectively. EDTA is a chelating agent, and seven trace elements are Iron chloride, Zinc chloride, Cobalt chloride, Sodium molybdate, calcium chloride, cuprous chloride, and Boric acid. The vitamins included in the experimental design solutions were Riboflavin, Panthothenic acid, nicotinic acid, Piridoxin, Biotin, and folic acid. In complex media, yeast extract ranges from 1 to 1.6 g $L^{-1,}$ is varied during preliminary concentrations. Screening designs often involve many factors and allow for initial differentiation of significant and non-significant factors and an estimation of the magnitude of the critical factors. A Full Factorial design, including 24 factors, would require almost 20 million experimental treatments. Fractional Factorial platform of the Design expert or JMP software can be used to generate 32 experimental treatments, randomly distributed into eight blocks. Each block comprised eight treatments and provided information on a technical error, a positive control, and negative control. These results are calculated using standard algorithms in the software.

E. coli growth was studied with nine continuous factors. The media ingredients in the first iteration were found to be influencing optical density in relation to time. For this study custom design platform of the Design expert software was used to construct a design that balanced the need to maximise the information that could be gathered from the experiment whilst minimising resources and time. Total number of experiments and concentration of yeast extract tested in 50–60 experiments is determined 10 g L^{-1} to be the optimal concentration [15].

3.2 Statistical designs for *E. coli* Media optimization

Typically, statistical medium optimisation is beneficial in improving overall product output, reduces time needed for process development and cost. The microbial processes have complex reactions. Evaluation of results statistically increases the reliability of results, further reduction in the number of experiments. In one of the study, the GDP mannose pyrophosphate yield was improved upto 100% after conducting 33 experiments [61]. Improvement of media by DOE is for understanding various test variants, multiple investigations, and uniform pattern. The results obtained after the various experiments are used to predict media improvement using mathematical models. The current advancement in statistical techniques provides rapid analysis of experimental findings. Meticulously planned experiments can enhance the desired outcomes using DOE strategies. For designing a full factorial, possible combinations of relevant factors, e.g. temperature, pH, buffers, carbon, nitrogen sources, strain, are considered. Similarly, partial factorial design is considered if knowledge about few components is not available. These experiments if planned and output is studied properly results in quick and definitive reproducible processes.

3.3 Identification of critical components: Plakett Burman design

Cultivation at a large scale requires a medium that will produce maximum yield of product per gram of substrate, maximum biomass and minimum undesirable by-products. Also, consisten with minimal problems during media preparation and sterilisation. While considering the biomass in isolation, it must be recognised that efficiently grown biomass produced by an optimised high productivity growth rate is not necessarily best suited for its ultimate purpose, such as synthesising the desired product. Different combination, and sequences of process condition need to be investigated to determine phases, specific sets of conditions during optimisation. OFAT for media optimisation using traditional replacement experiments with keeping one factor at a time for nutrient, antifoam, pH, temperature are highly time-consuming and expensive. Minimum number of experiments and development of process in short duration of time is prerequisite to media optimisation. Therefore, other alternative strategies must be considered, which allows more than one variable to be changed at a time, and these methods have been described in earlier studies by Placket and Burman 1946, and Hendrix 1980 (**Table 3**).

The Plakett Burman algorithm is a rapid statistical approach enables us to obtain the physicochemical parameters and factors influencing the fermentation process with the limited number of planned experiments [71]. For the given number of observation, the linear effect of all aspects are screened with maximum accuracy. The design is practical when investigating a large number of factors to produce an

		Factor 1	Factor 2	Factor 3	Factor 4	Factor 5	Factor 6
Std	Run	A:Select Soytone	B:Yeast Extract	C:Glucose	D:Buffer	E:Trace Metals	F:MgSO4
		g/L	g/L	g/L			g/L
6	1	0	0	0	Present	Absent	0.4
5	2	0	0	20	Absent	Present	0.4
9	3	16	24	20	Absent	Absent	0
11	4	16	0	20	Present	Present	0
1	5	16	24	0	Present	Present	0.4
8	6	16	24	0	Absent	Absent	0.4
10	7	0	24	20	Present	Absent	0
3	8	16	0	20	Present	Absent	0.4
7	9	16	0	0	Absent	Present	0
4	10	0	24	0	Present	Present	0
2	11	0	24	20	Absent	Present	0.4
12	12	0	0	0	Absent	Absent	0

Table 3.
Plackett-Burman design of experiments for media optimization (complex media example).

optimal or near-optimal response. Statistically optimised media design along with kinetic models characterises the fermentation behaviour more rapidly to achieve maximum productivity. Also, when complex carbon-nitrogen substrates, such as yeast extract or peptone, are used together with carbohydrate substrates, the Dissolved Oxygen (DO) change is not as significant when the carbon source is depleted, as cells continue to utilise the complex substrates [72]. The use of a good reliable model is essential to develop better strategies for optimising the fermentation process [73]. In one the study, during production of succinic acid [71] increasing output was achieved by combining Plackett-Burman design (PBD), steepest ascent method (SA), and Box–Behnken design (BBD) for fermentation medium. PBD identified Glucose, yeast extract, and $MgCO_3$ as critical components with optimal concentration was located to be 84.6 g L^{-1} of glucose, 14.5 g L^{-1} of yeast extract, and 64.7 g L^{-1} of $MgCO_3$ [2]. Also, the productivity was enhanced by 67.3% and 111.1%, respectively. Microbial fermentation for L-methionine (L-Met) production was enhanced by Plackett-Burman (PB) design, and Box–Behnken design (BBD) estimated glucose 37.43 g/L, yeast extract 0.95 g/L, KH_2PO_4 1.82 g/L, and $MgSO_4.7H_2O$ 4.51 g/L), L-Met titre was increased to 3.04 g/L from less than 2.0 g/L. an increase of 38.53% and 30.0% compared with those of the basal medium, respectively. Furthermore, higher L-Met productivity of 0.261 g/L/h was obtained, representing 2.13-fold higher in comparison to the original medium [14].

In another study, yield of O-succinyl-l-homoserine (OSH) was improved through multilevel fermentation optimisation; Plackett-Burman design was used to screen out three factors (glucose, yeast and threonine) from the original 11 factors that improved the titre of OSH.

Plackett Burman randomisation is an excellent tool for the determination of the effect of variables for optimisation. Once such approach for preparation of Bacterial Ghosts (BGs) preparation is established using these methods. The twelve experiments containing either the +1 or − 1 value for each variable in each experiment in random arrangement have been conducted simultaneously to get the best results and enable the best possible comparison. The BGQ has been given 100% quality as

10, while ten cells have been evaluated as either bad or good. This will decrease the range of the differences if we use %. Unexpectedly, *E. coli*, which is more sensitive to the SDS than *E. coli* BL21 (DE3), gives better results with most of the experiments. Nine experiments provide the number 10 out of the twelve experiments. Two give the number eight, and the only one shows the number 0, which means inferior preparation. The experimental Design is based mainly on the determination of Minimum Inhibition Concentration (MIC) and the Minimum Growth Concentration (MGC) of critical concentrations from chemical compounds able to convert viable cells to BGs. The mean of +1 experiments has been calculated using the following formula:$(\sum +\mathbf{1})/\boldsymbol{n}(+\mathbf{1})$. While the standard of −1 experiments has been calculated using the following formula: $(\sum -\mathbf{1})/\boldsymbol{n}(-\mathbf{1})$.The main effect of both +1 and − 1 for each variable has been calculated from the following formula: Main product = $\sum(+\mathbf{1})/\boldsymbol{n}(+\mathbf{1}) - \sum(-\mathbf{1})/\boldsymbol{n}(-\mathbf{1})$. Multiple linear regression analysis with ANOVA test of Plackett–Burman design has been performed on the BGQ as responses. A multiple linear regression analysis for the data of the BGQ has been committed to study the relationship between different variables and their level of significance regarding BGQ as a response. From the analysis of the Coefficient, Standard error, T Statistic, P-value and Confidence level % for each has been calculated. The confidence level has been calculated from the formula The confidence level% = 100∗(1 − P-value). The P-value from the ANOVA analysis for the BGQ response was determined to analyse the relationship between the variables at the 90% or higher confidence level. The model created from the analysis of Plackett–Burman experimental design using multiple regression analysis is based on the 1st order-model $\boldsymbol{Y} = \boldsymbol{\beta 0} + \sum \boldsymbol{\beta i X i}$. Where Y is the predicted response, ß0 model intercept, ßi variables linear coefficient. ANOVA test was generated for each response to determine the relationship between the variables at the 90% or higher confidence level [74]. This improvement was applied for optimisation of production of chimeric protein PfMSP3-MSP1_{19} resulted in critical concentrations are calculated are listed in **Table 4**.

3.4 Optimisation of fermentation conditions: Box-Behnken response surface methodology

George E. P. Box and Donald Behnken in 1960 developed the Box–Behnken response surface method. This algorithm establishes a comparison between composite central, three-level full factorial and Doehlert designs to optimise the fermentation conditions. In one of the example result of optimisation, the titre of O-succinyl-l-homoserine (OSH) reached 102.5 g l^{-1}, which is 5.6 times higher than before (15.6 g l^{-1}) [5]. Similarly, by Box–Behnken combination and Plackett-Burman design and were optimised further by employing the Response Surface Methodology, O-acetylhomoserine OAH production was up to 9.42 and 7.01 g/L. The effect of glycerol, ammonium chloride and yeast extract were screened for fermentation conditions [3].

scFv anti-HIV-1 P17 protein was optimised by the sequential simplex method. Plackett-Burman design (PBD) and sequential simplex were combined with the aim of improving feed medium for enhanced cell biomass, relative protein to biomass ratio. The scFv anti-p17 activity was enhanced by 4.43, 1.48, and 6.5 times more than batch cultivation, respectively [29].

DNA vaccine pcDNA-CCOL2A1 production was increased using the response surface method (RSM) in *E. coli* DH5alpha in fermentation, therapeutic DNA vaccine pcDNA-CCOL2A1 markedly increased from 223.37 mg/L to 339.32 mg/L under optimal conditions, and a 51.9% increase was observed compared with the original medium [21]. Statistical experimental design methodology for fermentation

Source of Variation	Degrees of Freedom	Sum of Squares [Partial]	Mean Squares [Partial]
Model	11	32.0001	2.9091
Main Effects	11	32.0001	2.9091
Residual	0	—	
Total	11	32.0001	
R-sq =	100.00%	—	
Regression Information		—	
Term	Effect	Coefficient	
Intercept		2.0933	
A:Select Soytone	1.5	0.75	
B:Yeast Extract	1.93	0.965	
C:Glucose	0.32	0.16	
D:Buffer	1.6467	0.8233	
E:Trace Metals	−0.4667	−0.2333	
F:MgSO4	−0.5733	−0.2867	
G:Dummy 1	−0.5667	−0.2833	
H:Dummy 2	0.6133	0.3067	
J:Dummy 3	0.3733	0.1867	
K:Dummy 4	−0.4467	−0.2233	
L:Dummy 5	−0.5433	−0.2717	

Table 4.
Plackett Burman design of experiments for media optimization after ANOVA analysis.

conditions (dissolved oxygen, IPTG, and temperature) improved rPDT production by *E. coli*. 15 Box–Behnken design augmented with centre points revealed that IPTG and DO at the centre point and low temperature would result in high yield. The optimal condition for rPDT production was found to be 100 mM IPTG, DO 30%, and temperature 20°C [23]. In another application, *E. coli* drug susceptibility testing was done by on-chip bacterial culture conditions using the Box–Behnken design response surface methodology for faster drug susceptibility, optimal growth parameters were determined within 6–8 h, MICs determination in 2–6 h of individual drugs (antibiotics and TCMs) to improve the clinical management of bacterial infection [75].

3.5 Functional characteristics with minimum experiments—Taguchi design

There are several challenges associated with the PBD and Box–Behnken design. To overcome these challenges of Box–Behnken new array based on "Orthogonal Array" was developed. Using this method, less number of experiments, instead of full factorial, is implemented. The system and technique provide control over three stages, likewise system strategy, tolerance design and parameters designing. The strategy design helps in determining tolerance, affecting factors in product output. Taguchi design is using a number of OAs to initiate the experimental setup, these arrays are utilised to suit the number of experimental iterations. Second step is conducting total tests with orthogonal arrays. These experiments are decided as per number of trial experiments as per Taguchi design, followed by randomisation of experiments for determining the output. This design analyses main effect and two-factor interactions. Noise for uncontrolled experimental variables is considered, focal point for two-point analysis. Taguchi methodology removes effect of noise due to uncontrolled variables; this is better as compared to PBD [76]. The Taguchi method provides help in functional characteristic for capturing acceptable deviations. Human insulin-like growth factor I (hIGF-I) was produced in one study in *E. coli*, 32y media, 32°C and 0.05 mM IPTG. The unimproved hIGF-I was 0.694 g L^{-1} which improved to 1.26 g L^{-1} using optimum conditions [77].

3.6 Deciphering outcome with- central composite design

This design is widely used in building a second-order (quadratic) model in response surface methodology (RSM). It consists of factorial Design with two levels +1 and − 1; centre points, factorial Design in experiments with median values; and star points identical runs for centre points except for one factor considering values above median and below the median. The number of star points is double the number of factors used. CCD is defined on the level of factors: as Face centred CCD (CCF), Inscribed CCD (CCI) and Circumcentered CCD (CCC). *E. coli* BL21(DE3) is utilised for optimum Design for extracellular production of recombinant human epidermal growth factor (rhEGF) by CCD. This resulted in 122.40 μg mL^{-1} rhEGF concentration in medium 20 h after induction. In 2 L fermentation, medium optimised yield to 1.5 fold and induction time to 3 h [78].

3.7 Predicting effects of responses—partial least squares modelling

The effect of media ingredients, interactions with the system (X) and co-relation in response to culture ΔOD_{600} (Y) is defined by the Partial least squares (PLS) model. PLS covariance of matrix design and outcome are inferred accurately by virtue of small underlying events not measured directly. The ideal or latent variable (LVs) to study outcome needs to be carefully evaluated to avoid overfitting training data. The prediction of the accuracy of models for DoE iterations with multiple values, the Root Mean Predicted Residual Error Sum of Squares, is with the lowest value of LVs, having the slightest error. The significance of LVs is calculated by the Voet T2 test. The score of media design and component determined by OD600, the threshold of 0.8 is accepted. The lower score of model threshold defines to remove from subsequent designs; these threshold values are considered to study the positive and negative effect on the contribution of various factors associated with the increasing growth of the culture. In one study, 2D spectrofluorometry was utilised for fermentation processes to monitor the fermentation process online to produce extracellular 5-aminolevulinic acid (ALA). Various chemometric methods used for analysis of the spectral data are principal component analysis (PCA), partial least square regression (PLS) and principal component regression (PCR). PCA results visualised and considered for online fermentation monitoring. PCR and PLS compared for correlation between the 2D fluorescence spectra, PLS had slightly better calibration and prediction performance than PCR [79].

3.8 Minimal product trial experiments—definitive screening designs

Traditional definitive screening designs (DSD) require a low number of experiments and trials to determine the positive outcome. Jones and Nachtscheim developed Jones DSD methodology. These designs are popular in biopharma due to the relatively small number of experiments. These designs use three levels for each factor, allows estimation of nonlinear effects. Evaluate the number of runs to determine X is 2 X + 1 or 2X + 3 for even and odd values. Typically, the Design of X = 6 is used if X < 6. There are few dummies runs with additional factorial or centre points are added to precisely determine the experimental error. In DSD, few different factorial trials or centre point trials are added to the initial design to define and evaluate the experimental error. One such example of DSD is for upstream process development for cell growth and increased product output in fed-batch high-cell-density fermentation. The expression of the desired gene cloned in the plasmid was under the control of the phoA promoter [13]. Simultaneous evaluation of phosphate concentration from 2.79 mM to 86.4 mM was designed using DoE. Several parameters, phosphate content, temperature, pH, and DO evaluated using a Definite screening design (DSD), resulting in determining each parameter's impact on product formation. Similarly, a 24-bioreactor ambr250™ system for fermentation utilised 10-factor DSD to characterise the process of demonstrating 16 batches reproducible workflow for recombinant protein production. This strategy was further evaluated by QbD approaches to assess techniques for late-stage depiction in small experiments and subsequently leading to large scale fermentation parameters improvement [80].

3.9 Stepwise regression and artificial neural network modelling

Artificial Neural Networks (ANN) are known for parallel, and continuous learning capabilities are known to interpret nonlinear functions. These are utilised to predict steady-state and dynamic processes. One iteration multi-layer perceptron (MLP) is famous for estimating hidden layers between output and input layers. Using this method, simulation of dissolve Oxygen (DO) parameters, Feeding (F), Biomass, Glucose, Acetate, and output production of γ-interferon is modelled. Several DoE iteration modelled using stepwise regression; these models are fitted with linear regression, six terms per model are allowed with Heredity Restriction. The goodness-of-fit of the resulting model is evaluated using Akaike Information Criterion (AICc). An artificial neural network (ANN) is used to create weighted ensemble of regression models. There are three nodes in single hidden layer of ANN. Sigmoid activation functions were used, cross-validation of 19 of the media formulations defined in the second DOE iteration were randomly selected and withheld from ANN training set to do validation studies [59].

4. DOE optimisation techniques for *E. coli*

4.1 Evaluation of production and process—response surface methodology (RSM)

RSM is simple, robust and efficient, in predicting processes of metabolite or product production. Also, this method helps in the determination of factors for specifications, changes in levels of the elements, response with specified levels, quantitative understanding of system behaviour, predict product properties, factor combinations not run and stability of the designed process.

RSM methodology consists of different phases. Typically, performed in three steps, First is the screening factors by steepest ascent/descent, secondly by quadratic regression model fitting, third optimization using canonical regression analysis.

For a cost-effective and robust process, improvising parameters related to medium, productivity, safety and usefulness are desired outputs. The interdependency of factors associated with productivity is difficult to understand, and this slows downs the enhancement process and yield evaluation.

Response surface methodology (RSM), is based on factorial designs to elevate the process and final product yield. RSM is considered a sturdy, robust, and efficient mathematical approach. It includes experimental statistical methods, multiple regression design, and analysis, resulting in developing the best strategies guided by constrained equations. RSM is typically applied to study the response of different media ingredients [21]. The production of Examples for *E. coli* fermentation. One such example is standardisation of production of human interferon-gamma (hIFN-γ).

$$Y = \beta_0 + \sum \beta_i x_i + \sum \beta_i i x^2 i_i + \sum \beta_i j x_i x_j + \varepsilon. \quad (3)$$

Where β_0 is defined as the constant, B_i the linear coefficient, $B_i i$ the quadratic coefficient and $B_i j$ the cross-product coefficient. X_i and X_j are levels of the independent variables, while ε is the residual error. This variable and RSM predicted 7.81 g L^{-1} glucose, 30°C for fermentation and induced at OD600 1.66, Combined with BBD to get the 95.50% acetate and 97.96% productivity of rhIFN-β [81].

Plackett Burman design and Response surface methodology are utilised together to increase the production of the desired product multiple-fold. Combining these techniques is usually employed to enhance the product outcome of several microbial processes, batch, fed-batch fermentation. RSM is widely used, with Plackett Burman, CCD, Box Benken. Even after much success, several limitations are associated with RSM, likewise predicting responses based on second-order polynomial equation [82], results in poor estimation of optimal designs, leading to low levels of yield or outcome. One limitation is developing a model for many variables on physical and chemical inputs due to nonlinear biochemical network interactions, with partial knowledge of these systems [83]. Another limitation is the study of multiple interactions and significant variations, resulting in error, bias and or no reproducibility. These challenges are dealt with better using Artificial neural networks (ANN) [84].

4.2 Study of interaction of pathways and multiple parameters—artificial neural network

An artificial neural network (ANN) is designed for a computing system to simulate the information and process the data similar to the human brain, guided by artificial intelligence (AI). It solves problems impossible or difficult by human or statistical standards. Handling units consist of inputs and outputs; using these inputs, ANNs produce desired or defined work. ANNs are built as neurons, are interconnected like a web in the human brain. There are hundreds or thousands of artificial neurons or processing units interconnected by nodes [85].

Similarly, as human functions, ANNs have a set of rules for learning backpropagation, an abbreviation for backward propagation of error, to perfect their output results. Typical processes in ANN are the training phase to recognise patterns in data, whether visually, aurally, or textually; in this supervised learning, the actual output is compared with the desired outcome. The differences are adjusted using backpropagation. The program runs backwards as we advance, and adjustments are made until the actual and expected output difference results in a minimum possible error. Designing of medium or metabolic process ANN is highly suitable, and it generates large amounts of data. The architecture of ANN consists of three layers: a layer of "input" connected to "hidden" units, ultimately connected with "output". The conditions for ANNs three types are Supervised, Unsupervised and Reinforced learning. The objective of supervised learning for the neural network is to provide input training data and possible experimental output. An unsupervised output unit is trained to respond to clusters or patterns present in the input data. Reinforced learning is an intermediate system; learning systems' actions are considered good or bad based on environmental responsibility. These parameters are adjusted till the time equilibrium state is attained. These systems are applied for system designing, modelling, optimisation. It leads to control the noisy signals and generalise through system training procedure. The ANNs are employed in various fermentation processes to optimise nutrient and prediction biomass outcome in-different culture conditions. ANN has several limitations; likewise, it needs proper training, also based on input data to get the quality output [86]. To overcome some of these challenges, ANN combined with a genetic algorithm (GA) is applied to improve the concentration and shelf life of aspartate-β-semialdehyde dehydrogenase protein [87].

4.3 Study of biological process of evolution- genetic algorithm (GA)

Genetic algorithm (GA), developed in 1975 by Holland and Long, is based on Charles Darwin's theory of natural selection. GA is a model for the study of biological evolution by testing crossover and recombination mutation in adaptive and artificial systems. A genetic algorithm works as a problem-solving strategy using essential genetic operators. There are several GA designed to deal with complex problems and parallelism for stationary or non-stationary functions, linear or nonlinear, continuous or discontinuous, random or noise. Improvement in yield of recombinant G-CSF was obtained in auto-induction medium. The backpropagation (BP) algorithm and radial basis function (RBF) algorithm combined with the Genetic Algorithm improve G-CSF yield. The yield of models was 72.24 and 76.09%, respectively, and are higher than those obtained using non-optimised auto-induction mediums.

There are some disadvantages or genetic algorithms as well. The formulation of the fitness function, population size, choice of factors for mutation and cross over, selection of criteria for these factors needed to be carried out carefully. Despite drawbacks, GA is one of the widely used algorithms in modern nonlinear optimisation [88].

4.4 Geometric function evaluation- Nelder-Mead simplex algorithm

Nelder–Mead published this algorithm in 1965. The objective is to solve the classical unconstrained optimisation problem of minimising a given nonlinear function; without derivative only numerical evaluation of the objective function is needed. This algorithm is based on geometry, and in three-dimensional space, simplex is a tetrahedron determined by four points (vertices) and their interconnecting line segments. For two dimensions, simplex is an equilateral triangle, and three dimensions should be tetrahedron. The objective function is evaluated every point with the highest numerical value of all four points is perpendicularly mirrored against the opposite plain segment, generating reflection [89]. An expansion can accompany the reflection to take more significant steps or a contraction to shrink the simplex where an optimisation valley floor is reached. The optimisation procedure continues until the termination criteria are met. The termination criterion is usually the maximum number of reflections with contractions or tolerance for optimisation variables. The algorithm can be implemented in N dimensions, where simplex is a hypercube with N + 1 vertex points. The NM method provides significant improvements in primary iteration and improves outcome. NM is combined with ANN to optimise the production of metabolites [90].

5. Problems and bottlenecks in *E. coli* media optimisation

Medium optimisation involves many experiments irrespective of the media chosen, which accounts for labour cost and is an open-ended experiment. Many experiments are carried out at shake flask, even after generation of large amount of data using single experiments. The results obtained at piolet scale batch fermentation are not reproducible. During shake flask experiments, the precise control over pH, oxygen transfer and evaporation is not controlled. The experiments carried out at shake flask may or may replicate during fermentation. Also, soluble proteins expression may lead to inclusion bodies formation. Optimisation of media is time consuming due to the requirement of rigorous experimental planning. Moreover, the media utilised in the production of recombinant products faces challenges due to variability in different batches, media availability, cost of media, bulk storage, transport time. For Biotherapeutics, Enzymes and Probiotics, the cost of media needs to lower in Probiotics compared to Enzymes and Biotherapeutics, respectively. The choice of fermentation depends on the solubility of protein from batch to fed-batch. *E. coli* cells are dynamic, and every product requires different media compared to the earlier optimised process. Optimisation of media depends on considering dynamic internal control mechanisms. For the production of metabolites after engineering of bacterial strains, metabolic pathways needed to be optimised to regulate the desired product by choice of media. The influence of using different strains for the production is dependent on toxicity, complexity (Disulphide bonds in the sequence), AT-rich sequences. In our previous study, *E. coli* cell Shuffle 3030H for production of *Plasmodium falciparum* MSP-3 and the MSP-1_{19} fusion protein was successfully optimised to generate protective antibodies [91].

Improvement of production of recombinant products is also guided by downstream processing of protein. Therefore, series of experiments designed for correct folding and confirmation are most important. Significant protein amounts can be achieved using pH, time for fermentation, oxygen transfer and temperature for fermentation. Also, inducer and harvest time are critical for increasing output. The critical factors for fermentation in batch and fed-batch are different. Therefore, the choice of media defined, semi-defined or complex media with vitamins, minerals and trace elements needed to be considered for evaluation in DOE experiments. To evaluate the considerable amount of output and variables combination and application of various algorithms is done to achieve desired output. In all the optimisation process and advanced algorithms such as Artificial neural networks and Genetic Algorithm are applied to achieve the desired output efficiently. The need for innovation as per sustainable development goals (SDGs) for United Nations 2030 plan is needed to increase the reach of technologies to low income countries. The application of DOE can improve the yield and cost leading to improved access to Biomolecules, Biopharmaceuticals, enzymes and metabolites.

6. Future strategies

The selection of host cells for industrial application has some technical difficulties despite the availability of many gene manipulations theoretically in various organisms. The availability of a genetic map, gene exchange system, useful vector and transformation procedures, and metabolic pathways leading from raw material to the desired product are essential criteria for selecting a suitable host strain. The most popular organisms used to date for the expression of the recombinant proteins are *E. coli*, *Bacillus subtilis*, *Bacillus stearothermophillus*, *Streptomyces spp*, *Corynebacterium*, *Saccharomyces cerevisiae*, *Pichia pastoris*, *Hansenula polymorpha* and various animal/plant cells. *E. coli* remains an important host system for the industrial protein production from cloned genes as one of the main applications of genetic engineering in biotechnology. Various efficient expression vector systems have been developed, and a variety of mutants are available as host strains for different purposes [92, 93]. Overexpression of a heterologous protein is possible in *E. coli*, making it suitable for industrial production. Fermentation DOE is an essential tool for basic research that greatly facilitates efficient purification and analysis of such proteins [94].

For the successful production of the recombinant protein-based vaccine, producing biologically active protein is an essential requirement that can be further scaled up. Production of a biologically active recombinant protein depends on the host cell's microenvironment for expression and compatibility of codon usage. *E. coli* has been a widely used expression host for the high-level production of heterologous protein. Differences in usage of codons in prokaryotes (*E. coli*) and eukaryotes Chinese hamster ovary cells (CHO) can substantially impact heterologous protein production. The compatibility of codon usage can significantly increase protein expression [95, 96]. Moreover, the presence of rare codons in cloned genes affects protein expression level and mRNA & plasmid stability. The excessive presence of rare codons may result in ribosome stalling, slow translation errors [96, 97]. In some cases, rare codons inhibit protein synthesis and cell growth [98]. Earlier studies of codon usage patterns in *E. coli* have established that a clear codon bias exists in the mRNA. The level of each cognate tRNA seems to be directly proportional to the codon frequency [99, 100]. The strategy widely used is to change rare codons in the target gene to the favoured codons of *E. coli* without affecting the encoded amino acid sequence [101, 102]. The second approach is to expand the intracellular tRNA pool by introducing a plasmid encoding additional copies of tRNAs for codons rarely used in *E. coli* [103]. The co-presence of the RIG plasmid encoding three tRNAsAG(A/G), ATA, GGA in the host cells significantly increases the expression level of Dihydropteroate synthase, Aldolase, Phosphatase, and Orotidine-5′-monophosphate decarboxylase of *P. falciparum* [104–106]. Codon optimisation for maximum expression of foreign proteins by changing host cell favourable codons is beneficial and crucial for large-scale proteins [107].

The recombinant plasmid carrying cloned gene would behave differently compared to the original vector plasmid. It can be easily understood, as it is preserved under a delicate quasi-equilibrium state in the host cell. There are several reasons for the instability of recombinants. The higher the plasmid gene expression, the more segregants (plasmid free cells) tend to appear. The recombinant plasmid is relatively unstable when the cloned gene products are inhibitory to the host cells. Phenotypic instability of plasmid is due to the disappearance of the entire plasmid or the deletion of a specific region [108]. Both plasmid copy number and plasmid loss rate are features affected by factors such as media composition growth rate and culture strategy [109] and other factors such as temperature, agitation rate, and pH [110].

Therefore, future strategies for optimisation of cultivations needed to be shift to conclusions evaluated during experimental phases before actual fermentation to identify role of batch, fed-batch or different media components. The utilisation of carbon, nitrogen and other minimally required nutrients during batch and fed-batch is critical for delivering output and achieve sustainable development goals (SDGs) for technological innovation. The method design and modelling approaches are future strategies for increasing output during a process development. Utilisation of one factor and carrying out experiments by statistical media optimisation can be improved by combining several algorithms such as Plakett Burman, Box–Behnken, Taguchi design, Central composite design, partial least squares modelling in determining optimal factors. Response surface methodology with Artificial neural network (ANNs) can be applied to difficult model kind of fermentation processes. A free artificial neural network is applicable for carrying out nonlinear regression models to optimise metabolic processes. These algorithms are combined and applied to increase productivity and optimise the product output by reducing the cost of fermentation and product development.

7. Conclusions

Optimisation of critical factors and nutrient sources is an essential step for metabolite, recombinant proteins before pilot fermentation. In this chapter, strategies, conventional, advanced process design are reviewed and detailed. DOE approaches with statistical evaluation are critical for process development are essential for saving experimentation time. The strategies and examples shared in this review have been analysed for ease of implementation, time consumption. The conditions and media designed needs to be further tested under realistic conditions, full scale process with replication to production setup.

Overall, this chapter detailed need of critical factors identification, their significant contribution in enhancing process of metabolite production. Also, recently, co-fermentation of glycerol and glucose in engineered *E. coli* increased production of 1, 3 propanediol [1]. Similarly, O-acetylhomoserine production is increased by suitable designs for fermentation and modification of glycerol-Oxidative pathway [3]. Production of recombinant protein in one study response surface methodology was utilised for production of repletase and improved yield to 188 mg $L^{-1 \text{ of}}$ fermentation [15]. The approaches discussed in this chapter have several advantages for improving the yield and reduction of resource utilisation. These approaches are efficient for achieving the access for biotechnologically produced products to reach the larger population across the globe.

Acknowledgements

Puneet Kumar Gupta acknowledges financial support from Grace Lifetech Pvt. Ltd. for the study. Jyotheeswara Reddy Edula acknowledges support from the Tata Institute of Genomics and Society (TIGS).

Conflict of interest

The authors declare no conflict of interest.

Acronyms and Abbreviations

DOE	Design of Experiments
E. coli	*Escherichia coli*
OFAT	one-factor-at-a-time
ANOVA	Analysis of variance
RSM	Response surface methodology

DNA	Deoxyribonucleic acid
RNA	Ribonucleic Acid
IPTG	Isopropyl β-D-1-thiogalactopyranoside
EDTA	Ethylenediamine tetra acetic acid
DO	Dissolved Oxygen
PBD	Plackett-Burman design
SA	Steepest ascent method
BBD	Box–Behnken design
OSH	O-succinyl-l-homoserine
BGs	Bacterial Ghosts
MIC	Minimum Inhibition Concentration
MGC	Minimum Growth Concentration
BGQ	Bacterial Ghosts Quantification
OA	Orthogonal Array
CCF	Face centred CCD
CCI	Inscribed CCD
CCC	Circumcentered CCD
DSD	Definitive screening designs
QbD	Quality by design
ANN	Artificial Neural Network
GA	Genetic Algorithm
NM	Nelder–Mead
SDGs	Sustainable Development Goals
CHO	Chinese Hamster Ovary

Author details

Puneet Kumar Gupta[1*] and Jyotheeswara Reddy Edula[2]

1 Grace Lifetech Pvt. Ltd., New Delhi, India

2 Tata Institute of Genetics and Society, Bangalore, Karnataka, India

*Address all correspondence to: guptapuneetin@gmail.com

References

[1] Yun J, Zabed HM, Zhang Y, Parvez A, Zhang G, Qi X. Co-fermentation of glycerol and glucose by a co-culture system of engineered *Escherichia coli* strains for 1,3-propanediol production without vitamin B12 supplementation. Bioresour Technol. 2020/10/14 ed2021. p. 124218.

[2] Xu JM, Cao HT, Wang M, Ma BJ, Wang LY, Zhang K, et al. 2021. Development of a Combination Fermentation Strategy to Simultaneously Increase Biomass and Enzyme Activity of D-amino Acid Oxidase Expressed in Escherichia coli. Appl Biochem Biotechnol. DOI: 10.1007/s12010-021-03519-7.

[3] Liu P, Liu JS, Zhang B, Liu ZQ, Zheng YG. 2021. Increasement of O-acetylhomoserine production in Escherichia coli by modification of glycerol-oxidative pathway coupled with optimisation of fermentation. Biotechnol Lett 43 (1):105-117. DOI: 10.1007/s10529-020-03031-8.

[4] Datta P, Fu L, Brodfuerer P, Dordick JS, Linhardt RJ. 2021. High density fermentation of probiotic E. coli Nissle 1917 towards heparosan production, characterisation, and modification. Appl Microbiol Biotechnol 105 (3):1051-1062. DOI: 10.1007/s00253-020-11079-9.

[5] Zhu WY, Niu K, Liu P, Fan YH, Liu ZQ, Zheng YG. 2020. Enhanced O-succinyl-l-homoserine production by recombinant Escherichia coli DeltaIJBB*TrcmetL/pTrc-metA(fbr) -Trc-thrA(fbr) -yjeH via multilevel fermentation optimization. J Appl Microbiol. DOI: 10.1111/jam.14884.

[6] Zhou S, Hao T, Zhou J. 2020. Fermentation and Metabolic Pathway Optimization to De Novo Synthesize (2S)-Naringenin in Escherichia coli. J Microbiol Biotechnol 30 (10):1574-1582. DOI: 10.4014/jmb.2008.08005.

[7] Zhao C, Fang H, Wang J, Zhang S, Zhao X, Li Z, et al. 2020. Application of fermentation process control to increase l-tryptophan production in Escherichia coli. Biotechnol Prog 36 (2):e2944. DOI: 10.1002/btpr.2944.

[8] Yao P, You S, Qi W, Su R, He Z. 2020. Investigation of fermentation conditions of biodiesel by-products for high production of beta-farnesene by an engineered Escherichia coli. Environ Sci Pollut Res Int 27 (18):22758-22769. DOI: 10.1007/s11356-020-08893-z.

[9] Xu D, Zhang Z, Liu Z, Xu Q. 2020. Using enzymatic hydrolyzate as new nitrogen source for L-tryptophan fermentation by E.coli. Bioengineered 11 (1):1-10. DOI: 10.1080/21655979.2019.1700092.

[10] Wang J, Liu Y, Chen Y, Wang A, Wei Q, Liu D, et al. 2020. Large-scale manufacture of VP2 VLP vaccine against porcine parvovirus in Escherichia coli with high-density fermentation. Appl Microbiol Biotechnol 104 (9):3847-3857. DOI: 10.1007/s00253-020-10483-5.

[11] Li D, Fang H, Gai Y, Zhao J, Jiang P, Wang L, et al. 2020. Metabolic engineering and optimisation of the fermentation medium for vitamin B12 production in Escherichia coli. Bioprocess Biosyst Eng 43 (10):1735-1745. DOI: 10.1007/s00449-020-02355-z.

[12] Hao Y, Ma Q, Liu X, Fan X, Men J, Wu H, et al. 2020. High-yield production of L-valine in engineered Escherichia coli by a novel two-stage fermentation. Metab Eng 62:198-206. DOI: 10.1016/j.ymben.2020.09.007.

[13] Agbogbo FK, Ramsey P, George R, Joy J, Srivastava S, Huang M, et al. 2020.

Upstream development of Escherichia coli fermentation process with PhoA promotdesignng design of experiments (DoE). J Ind Microbiol Biotechnol 47 (9–10):789-799. DOI: 10.1007/s10295-020-02302-7.

[14] Zhou HY, Wu WJ, Niu K, Xu YY, Liu ZQ, Zheng YG. 2019. Enhanced L-methionine production by genetically engineered Escherichia coli through fermentation optimisation. 3 Biotech 9 (3):96. DOI: 10.1007/s13205-019-1609-8.

[15] Zare H, Mir Mohammad Sadeghi H, Akbari V. 2019. Optimisation of Fermentation Conditions for Reteplase Expression by Escherichia coli Using Response Surface Methodology. Avicenna J Med Biotechnol 11 (2):162-168.

[16] Xu Y, Yang L, Zhao S, Wang Z. 2019. Large-scale production of tauroursodeoxycholic acid products through fermentation optimisation of engineered Escherichia coli cell factory. Microb Cell Fact 18 (1):34. DOI: 10.1186/s12934-019-1076-2.

[17] Tang XL, Wang ZC, Yang J, Zheng RC, Zheng YG. 2019. Statistical medium optimisation and DO-STAT fed-batch fermentation for enhanced production of tyrosine phenol lyase in recombinant Escherichia coli. Prep Biochem Biotechnol 49 (2):117-126. DOI: 10.1080/10826068.2018.1541808.

[18] Horner M, Gerhardt K, Salavei P, Hoess P, Harrer D, Kaiser J, et al. 2019. Production of Phytochromes by High-Cell-Density E. coli Fermentation. ACS Synth Biol 8 (10):2442-2450. DOI: 10.1021/acssynbio.9b00267.

[19] Fang Z, Sha C, Peng Z, Zhang J, Du G. 2019. Protein engineering to enhance keratinolytic protease activity and excretion in Escherichia coli and its scale-up fermentation for high extracellular yield. Enzyme Microb Technol 121:37-44. DOI: 10.1016/j.enzmictec.2018.11.003.

[20] Chen T, Liu N, Ren P, Xi X, Yang L, Sun W, et al. 2019. Efficient Biofilm-Based Fermentation Strategies for L-Threonine Production by Escherichia coli. Front Microbiol 10:1773. DOI: 10.3389/fmicb.2019.01773.

[21] Long J, Zhao X, Liang F, Liu N, Sun Y, Xi Y. 2018. Optimisation of fermentation conditions for an Escherichia coli strain engineered using the response surface method to produce a novel therapeutic DNA vaccine for rheumatoid arthritis. J Biol Eng 12:22. DOI: 10.1186/s13036-018-0110-y.

[22] Liang J, Zhao J, Wang Z, Wang Y. 2018. Temperature gradient-based high-cell density fed-batch fermentation for the production of pyruvate oxidase by recombinant E. coli. Prep Biochem Biotechnol 48 (2):188-193. DOI: 10.1080/10826068.2018.1425709.

[23] Koopaei NN, Khadiv-Parsi P, Khoshayand MR, Mazlomi MA, Kebriaeezadeh A, Moloudian H, et al. 2018. Optimisation of rPDT fusion protein expression by Escherichia coli in pilot scale fermentation: a statistical experimental design approach. AMB Express 8 (1):135. DOI: 10.1186/s13568-018-0667-3.

[24] Jing J, Chen Y, Sheng L, Wu M. 2018. Optimised production of insulin variant, a recombinant platelet aggregation inhibitor, by high cell-density fermentation of recombinant Escherichia coli. Protein Expr Purif 152: 7-12. DOI: 10.1016/j.pep.2018.07.001.

[25] Sar T, Seker G, Erman AG, Stark BC, Yesilcimen Akbas M. 2017. Repeated batch fermentation of immobilised E. coli expressing Vitreoscilla hemoglobin for long-term use. Bioengineered 8 (5):651-660. DOI: 10.1080/21655979.2017.1303024.

[26] Fu X, Wang Y, Wang J, Garza E, Manow R, Zhou S. 2017. Semi-industrial scale (30 m(3)) fed-batch fermentation for the production of D-lactate by

Escherichia coli strain HBUT-D15. J Ind Microbiol Biotechnol 44 (2):221-228. DOI: 10.1007/s10295-016-1877-9.

[27] Couto MR, Rodrigues JL, Rodrigues LR. 2017. Optimisation of fermentation conditions for the production of curcumin by engineered Escherichia coli. J R Soc Interface 14 (133). DOI: 10.1098/rsif.2017.0470.

[28] Zheng H, Yu Z, Fu X, Li S, Xu J, Song H, et al. 2016. High level extracellular production of a truncated alkaline beta-mannanase from alkaliphilic Bacillus sp. N16-5 in Escherichia coli by the optimisation of induction condition and fed-batch fermentation. J Ind Microbiol Biotechnol 43 (7):977-987. DOI: 10.1007/s10295-016-1773-3.

[29] Paopang P, Kasinrerk W, Tayapiwatana C, Seesuriyachan P, Butr-Indr B. 2016. Multiparameter optimisation method and enhanced production of secreted recombinant single-chain variable fragment against the HIV-1 P17 protein from Escherichia coli by fed-batch fermentation. Prep Biochem Biotechnol 46 (3):305-312. DOI: 10.1080/10826068.2015.1031388.

[30] Lu XX, Yi Y, Su QD, Bi SL. 2016. Expression and Purification of Recombinant Hepatitis Delta Virus (HDV) Antigen for Use in a Diagnostic ELISA for HDV Infection Using the High-Density Fermentation Strategy in Escherichia coli. Biomed Environ Sci 29 (6):417-423. DOI: 10.3967/bes2016.054.

[31] Huang L, Wang Q, Jiang S, Zhou Y, Zhang G, Ma Y. 2016. Improved extracellular expression and high-cell-density fed-batch fermentation of chitosanase from Aspergillus Fumigatus in Escherichia coli. Bioprocess Biosyst Eng 39 (11):1679-1687. DOI: 10.1007/s00449-016-1643-4.

[32] Sohoni SV, Nelapati D, Sathe S, Javadekar-Subhedar V, Gaikaiwari RP, Wangikar PP. 2015. Optimisation of high cell density fermentation process for recombinant nitrilase production in E. coli. Bioresour Technol 188: 202-208. DOI: 10.1016/j.biortech.2015.02.038.

[33] Morowvat MH, Babaeipour V, Rajabi Memari H, Vahidi H. 2015. Optimisation of Fermentation Conditions for Recombinant Human Interferon Beta Production by Escherichia coli Using the Response Surface Methodology. Jundishapur J Microbiol 8 (4):e16236. DOI: 10.5812/jjm.8(4)2015.16236.

[34] Minamoto T, Takahashi N, Kitahara S, Shinozaki Y, Hirano T, Hakamata W, et al. 2015. Saccharification of beta-chitin from squid pen by a fermentation method using recombinant chitinase-secreting Escherichia coli. Appl Biochem Biotechnol 175 (8):3788-3799. DOI: 10.1007/s12010-015-1547-9.

[35] Munoz-Gutierrez I, Moss-Acosta C, Trujillo-Martinez B, Gosset G, Martinez A. 2014. Ag43-mediated display of a thermostable beta-glucosidase in Escherichia coli and its use for simultaneous saccharification and fermentation at high temperatures. Microb Cell Fact 13:106. DOI: 10.1186/s12934-014-0106-3.

[36] Li Q, Mannall GJ, Ali S, Hoare M. 2013. An ultra scale-down approach to study the interaction of fermentation, homogenisation, and centrifugation for antibody fragment recovery from rec E. coli. Biotechnol Bioeng 110 (8): 2150-2160. DOI: 10.1002/bit.24891.

[37] Xu J, Qian Y, Skonezny PM, You L, Xing Z, Meyers DS, et al. 2012. Reduction of N-terminal methionylation while increasing titer by lowering metabolic and protein production rates in E. coli auto-induced fed-batch fermentation. J Ind Microbiol Biotechnol 39 (8):1199-1208. DOI: 10.1007/s10295-012-1127-8.

[38] Silva R, Ferreira S, Bonifacio MJ, Dias JM, Queiroz JA, Passarinha LA. 2012. Optimisation of fermentation conditions for the production of human soluble catechol-O-methyltransferase by Escherichia coli using artificial neural network. J Biotechnol 160 (3–4): 161-168. DOI: 10.1016/j.jbiotec.2012.03.025.

[39] Seo MJ, Choi HJ, Chung KH, Pyun YR. 2011. Production of a platelet aggregation inhibitor, salmosin, by high cell density fermentation of recombinant Escherichia coli. J Microbiol Biotechnol 21 (10):1053-1056. DOI: 10.4014/jmb.1105.05014.

[40] Premsukh A, Lavoie JM, Cizeau J, Entwistle J, MacDonald GC. 2011. Development of a GMP Phase III purification process for VB4-845, an immunotoxin expressed in E. coli using high cell density fermentation. Protein Expr Purif 78 (1):27-37. DOI: 10.1016/j.pep.2011.03.009.

[41] Honghong J, Fuping L, Yu L, Xiaoguang L, Yihan L, Hongbin W, et al. 2011. Synthesis of GDP-mannose using coupling fermentation of recombinant Escherichia coli. Biotechnol Lett 33 (6): 1145-1150. DOI: 10.1007/s10529-011-0547-2.

[42] Wang Z, Ly M, Zhang F, Zhong W, Suen A, Hickey AM, et al. 2010. E. coli K5 fermentation and the preparation of heparosan, a bioengineered heparin precursor. Biotechnol Bioeng 107 (6): 964-973. DOI: 10.1002/bit.22898.

[43] Brusehaber E, Schwiebs A, Schmidt M, Bottcher D, Bornscheuer UT. 2010. Production of pig liver esterase in batch fermentation of E. coli Origami. Appl Microbiol Biotechnol 86 (5):1337-1344. DOI: 10.1007/s00253-009-2392-y.

[44] Zhang D, Wang PG, Qi Q. 2007. A two-step fermentation process for efficient production of penta-N-acetyl-chitopentaose in recombinant Escherichia coli. Biotechnol Lett 29 (11): 1729-1733. DOI: 10.1007/s10529-007-9462-y.

[45] Kim M, Elvin C, Brownlee A, Lyons R. 2007. High yield expression of recombinant pro-resilin: lactose-induced fermentation in E. coli and facile purification. Protein Expr Purif 52 (1):230-236. DOI: 10.1016/j.pep.2006.11.003.

[46] Yazdani SS, Shakri AR, Chitnis CE. 2004. A high cell density fermentation strategy to produce recombinant malarial antigen in E. coli. Biotechnol Lett 26 (24):1891-1895. DOI: 10.1007/s10529-004-6040-4.

[47] Kleist S, Miksch G, Hitzmann B, Arndt M, Friehs K, Flaschel E. 2003. Optimisation of the extracellular production of a bacterial phytase with Escherichia coli by using different fed-batch fermentation strategies. Appl Microbiol Biotechnol 61 (5–6):456-462. DOI: 10.1007/s00253-003-1229-3.

[48] Khalilzadeh R, Shojaosadati SA, Bahrami A, Maghsoudi N. 2003. Over-expression of recombinant human interferon-gamma in high cell density fermentation of Escherichia coli. Biotechnol Lett 25 (23):1989-1992. DOI: 10.1023/b:bile.0000004390.98648.25.

[49] Panda AK, Ghorpade A, Mukhopadhyay A, Talwar GP, Garg LC. 1995. High cell density fermentation of recombinant Vibrio cholerae for the production of B subunit of Escherichia coli enterotoxin. Biotechnol Bioeng 45 (3):245-250. DOI: 10.1002/bit.260450309.

[50] Sakamoto S, Terada I, Iijima M, Matsuzawa H, Ohta T. 1994. Fermentation conditions for efficient production of thermophilic protease in Escherichia coli harboring a plasmid. Appl Microbiol Biotechnol 42 (4): 569-574. DOI: 10.1007/BF00173922.

[51] Kapralek F, Jecmen P, Sedlacek J, Fabry M, Zadrazil S. 1991. Fermentation conditions for high-level expression of the tac-promoter-controlled calf prochymosin cDNA in Escherichia coli HB101. Biotechnol Bioeng 37 (1):71-79. DOI: 10.1002/bit.260370111.

[52] Herber WK, Bailey FJ, Carty CE, Heimbach JC, Maigetter RZ. 1991. Production of cytotoxic proteins in Escherichia coli: a fermentation process for producing enzymatically active HIV-1 protease. Appl Microbiol Biotechnol 36 (2):149-152. DOI: 10.1007/BF00164410.

[53] Chang LL, Hwang LY, Hwang CF, Mou DG. 1991. Study of high density Escherichia coli fermentation for production of porcine somatotropin protein. Ann N Y Acad Sci 646: 259-272. DOI: 10.1111/j.1749-6632.1991.tb18589.x.

[54] Riesenberg D, Menzel K, Schulz V, Schumann K, Veith G, Zuber G, et al. 1990. High cell density fermentation of recombinant Escherichia coli expressing human interferon alpha 1. Appl Microbiol Biotechnol 34 (1):77-82. DOI: 10.1007/BF00170927.

[55] Jung G, Denefle P, Becquart J, Mayaux JF. 1988. High-cell density fermentation studies of recombinant Escherichia coli strains expressing human interleukin-1 beta. Ann Inst Pasteur Microbiol 139 (1):129-146. DOI: 10.1016/0769-2609(88)90100-7.

[56] Botterman JH, De Buyser DR, Spriet JA, Zabeau M, Vansteenkiste GC. 1985. Fermentation and recovery of the EcoRI restriction enzyme with a genetically modified Escherichia coli strain. Biotechnol Bioeng 27 (9): 1320-1327. DOI: 10.1002/bit.260270908.

[57] Zhang J, Greasham R. 1999. Chemically defined media for commercial fermentations. Applied microbiology and biotechnology 51 (4): 407-421. DOI: 10.1007/s002530051411.

[58] Nikel PI, Pettinari MJ, Galvagno MA, Méndez BS. 2008. Poly(3-hydroxybutyrate) synthesis from glycerol by a recombinant Escherichia coli arcA mutant in fed-batch microaerobic cultures. Applied microbiology and biotechnology 77 (6):1337-1343. DOI: 10.1007/s00253-007-1255-7.

[59] Singleton C, Gilman J, Rollit J, Zhang K, Parker DA, Love J. 2019. A design of experiments approach for the rapid formulation of a chemically defined medium for metabolic profiling of industrially important microbes. PLoS One 14 (6):e0218208. DOI: 10.1371/journal.pone.0218208.

[60] Singh V, Haque S, Niwas R, Srivastava A, Pasupuleti M, Tripathi CK. 2016. Strategies for Fermentation Medium Optimisation: An In-Depth Review. Front Microbiol 7:2087. DOI: 10.3389/fmicb.2016.02087.

[61] Weuster-Botz D. 2000. Expedesignal design for fermentation media development: Stadesignal design or global random search? Journal of Bioscience and Bioengineering 90 (5): 473–83. DOI: https://doi.org/10.1016/S1389-1723(01)80027-X.

[62] Weissman SA, Anderson NG. 2015. Design of Experiments (DoE) and Process Optimisation. A Review of Recent Publications. Organic Process Research & Development 19 (11): 1605-1633. DOI: 10.1021/op500169m.

[63] Dvorak P, Chrast L, Nikel PI, Fedr R, Soucek K, Sedlackova M, et al. 2015. Exacerbation of substrate toxicity by IPTG in Escherichia coli BL21(DE3) carrying a synthetic metabolic pathway. Microbial Cell Factories 14 (1):201. DOI: 10.1186/s12934-015-0393-3.

[64] Cardoso VM, Campani G, Santos MP, Silva GG, Pires MC, Gonçalves VM, et al. 2020. Cost analysis based on bioreactor cultivation conditions: Production of a soluble

recombinant protein using Escherichia coli BL21(DE3). Biotechnology Reports 26:e00441. DOI: https://doi.org/10.1016/j.btre.2020.e00441.

[65] Siegele DA, Hu JC. 1997. Gene expression from plasmids containing the <em>araBAD</em> promoter at subsaturating inducer concentrations represents mixed populations. Proceedings of the National Academy of Sciences 94 (15):8168-8172. DOI: 10.1073/pnas.94.15.8168.

[66] Choi YJ, Morel L, Le François T, Bourque D, Bourget L, Groleau D, et al. 2010. Novel, versatile, and tightly regulated expression system for Escherichia coli strains. Applied and environmental microbiology 76 (15): 5058-5066. DOI: 10.1128/AEM.00413-10.

[67] Mujacic M, Cooper KW, Baneyx F. 1999. Cold-inducible cloning vectors for low-temperature protein expression in Escherichia coli: application to the production of a toxic and proteolytically sensitive fusion protein. Gene 238 (2): 325-332. DOI: 10.1016/s0378-1119(99)00328-5.

[68] Nor NM, Mohamed MS, Loh TC, Foo HL, Rahim RA, Tan JS, et al. 2017. Comparative analyses on medium optimisation using one-factor-at-a-time, response surface methodology, and artificial neural network for lysine–methionine biosynthesis by Pediococcus pentosaceus RF-1. Biotechnology & Biotechnological Equipment 31 (5): 935-947. DOI: 10.1080/13102818.2017.1335177.

[69] Elbing KL, Brent R. 2019. Recipes and Tools for Culture of Escherichia coli. Current protocols in molecular biology 125 (1):e83-e. DOI: 10.1002/cpmb.83.

[70] Zhang C, Chen X, Zou R, Zhou K, Stephanopoulos G, Too H-P. 2013. Combining Genotype Improvement and Statistical Media Optimization for Isoprenoid Production in E. coli. PLoS One 8 (10):e75164. DOI: 10.1371/journal.pone.0075164.

[71] Zhu LW, Wang CC, Liu RS, Li HM, Wan DJ, Tang YJ. 2012. Actinobacillus succinogenes ATCC 55618 fermentation medium optimisation for the production of succinic acid by response surface methodology. J Biomed Biotechnol 2012: 626137. DOI: 10.1155/2012/626137.

[72] Riesenberg D, Schulz V, Knorre WA, Pohl HD, Korz D, Sanders EA, et al. 1991. High cell density cultivation of Escherichia coli at controlled specific growth rate. J Biotechnol 20 (1):17-27. DOI: 10.1016/0168-1656(91)90032-Q.

[73] Ghaly AE, Kamal M, Correia LR. 2005. Kinetic modelling of continuous submerged fermentation of cheese whey for single cell protein production. Bioresour Technol 96 (10):1143-1152. DOI: 10.1016/j.biortech.2004.09.027.

[74] Amro AA, Salem-Bekhit MM, Alanazi FK. 2014. Plackett-Burman randomisation method for Bacterial Ghosts preparation form E. coli JM109. Saudi Pharm J 22 (3):273-279. DOI: 10.1016/j.jsps.2013.06.002.

[75] Li H, van den Driesche S, Bunge F, Yang B, Vellekoop MJ. 2019. Optimisation of on-chip bacterial culture conditions using the Box-Behnken design response surface methodology for faster drug susceptibility screening. Talanta 194: 627-33. DOI: https://doi.org/10.1016/j.talanta.2018.10.048.

[76] Aggarwal A, Singh H. 2005. Optimisation of machining techniques — A retrospective and literature review. Sadhana 30 (6):699-711. DOI: 10.1007/BF02716704.

[77] Ranjbari J, Babaeipour V, Vahidi H, Moghimi H, Mofid MR, Namvaran MM, et al. 2015. Enhanced Production of Insulin-like Growth Factor I Protein in Escherichia coli by Optimisation of Five

Key Factors. Iran J Pharm Res 14 (3): 907-917.

[78] Sriwidodo S, Subroto T, Maksum I, Wathoni N, Rostinawati T, Ulya H, et al. 2019. Optimisation of secreted recombinant human epidermal growth factor production using pectate lyase B from *Escherichia coli* BL21(DE3) by central composite design and its production in high cell density culture. Journal of Pharmacy And Bioallied Sciences 11 (8):562-566. DOI: 10.4103/jpbs.JPBS_207_19.

[79] Rhee JI, Kang T-H. 2007. On-line process monitoring and chemometric modeling with 2D fluorescence spectra obtained in recombinant E. coli fermentations. Process Biochemistry 42 (7):1124-34. DOI: https://doi.org/10.1016/j.procbio.2007.05.007.

[80] Tai M, Ly A, Leung I, Nayar G. 2015. Efficient high-throughput biological process characterisation: Definitive screening design with the ambr250 bioreactor system. Biotechnol Prog 31 (5):1388-1395. DOI: 10.1002/btpr.2142.

[81] Feyzdar M, Vali AR, Babaeipour V. 2013. Identification and Optimisation of Recombinant E. coli Fed-Batch Fermentation Producing γ-Interferon Protein. International Journal of Chemical Reactor Engineering 11 (1): 123-134. DOI: doi:10.1515/ijcre-2012-0081.

[82] Zhang C, Fan D, Shang La, Ma X, Luo Ye, Xue W, et al. 2010. Optimisation of Fermentation Process for Human-like Collagen Production of Recombinant Escherichia coli Using Response Surface Methodology. Chinese Journal of Chemical Engineering 18 (1): 137-42. DOI: https://doi.org/10.1016/S1004-9541(08)60334-1.

[83] Aydar AY. 2018. Utilisation of response surface methodology in optimisation of extraction of plant materials. Statistical approaches with emphasis on design of experiments applied to chemical processes:157-169.

[84] Desai KM, Survase SA, Saudagar PS, Lele SS, Singhal RS. 2008. Comparison of artificial neural network (ANN) and response surface methodology (RSM) in fermentation media optimisation: Case study of fermentative production of scleroglucan. Biochemical Engineering Journal 41 (3):266-73. DOI: https://doi.org/10.1016/j.bej.2008.05.009.

[85] Kubilius J, Schrimpf M, Kar K, Hong H, Majaj NJ, Rajalingham R, et al. 2019. Brain-like object recognition with high-performing shallow recurrent ANNs. arXiv preprint arXiv:190906161.

[86] Mishra M, Srivastava M, editors. A view of artificial neural network. 2014 International Conference on Advances in Engineering & Technology Research (ICAETR-2014); 2014: IEEE.

[87] Khan S, Bhakuni V, Praveen V, Tewari R, Tripathi CKM, Gupta VD. 2011. Maximising the native concentration and shelf life of protein: a multiobjective optimisation to reduce aggregation. Applied microbiology and biotechnology 89 (1):99-108. DOI: 10.1007/s00253-010-2835-5.

[88] Beg AH, Islam MZ, editors. Advantages and limitations of genetic algorithms for clustering records. 2016 IEEE 11th Conference on Industrial Electronics and Applications (ICIEA); 2016: IEEE.

[89] Lagarias JC, Poonen B, Wright MH. 2012. Convergence of the Restricted Nelder–Mead Algorithm in Two Dimensions. SIAM Journal on Optimization 22 (2):501-532.

[90] Singer S, Nelder J. 2009. Nelder-mead algorithm. Scholarpedia 4 (7):2928.

[91] Gupta PK, Mukherjee P, Dhawan S, Pandey AK, Mazumdar S, Gaur D, et al.

2014. Production and preclinical evaluation of Plasmodium falciparum MSP-119 and MSP-311 chimeric protein, PfMSP-Fu24. Clinical and vaccine immunology : CVI. DOI: 10.1128/CVI.00179-14.

[92] Favre-Bulle O, Schouten T, Kingma J, Witholt B. 1991. Bioconversion of n-octane to octanoic acid by a recombinant Escherichia coli cultured in a two-liquid phase bioreactor. Biotechnology (N Y) 9 (4): 367-371.

[93] Sabatie J, Speck D, Reymund J, Hebert C, Caussin L, Weltin D, et al. 1991. Biotin formation by recombinant strains of Escherichia coli: influence of the host physiology. J Biotechnol 20 (1): 29-49.

[94] Wagner CR, Benkovic SJ. 1990. Site directed mutagenesis: a tool for enzyme mechanism dissection. Trends Biotechnol 8 (9):263-270.

[95] Kane JF. 1995. Effects of rare codon clusters on high-level expression of heterologous proteins in Escherichia coli. Curr Opin Biotechnol 6 (5): 494-500.

[96] Kurland C, Gallant J. 1996. Errors of heterologous protein expression. Curr Opin Biotechnol 7 (5):489-493.

[97] Roche ED, Sauer RT. 1999. SsrA-mediated peptide tagging caused by rare codons and tRNA scarcity. EMBO J 18 (16):4579-4589. DOI: 10.1093/emboj/18.16.4579.

[98] Zahn K. 1996. Overexpression of an mRNA dependent on rare codons inhibits protein synthesis and cell growth. J Bacteriol 178 (10): 2926-2933.

[99] Gouy M, Gautier C. 1982. Codon usage in bacteria: correlation with gene expressivity. Nucleic Acids Res 10 (22): 7055-7074.

[100] Zhang SP, Zubay G, Goldman E. 1991. Low-usage codons in Escherichia coli, yeast, fruit fly and primates. Gene 105 (1):61-72.

[101] Prapunwattana P, Sirawaraporn W, Yuthavong Y, Santi DV. 1996. Chemical synthesis of the Plasmodium falciparum dihydrofolate reductase-thymidylate synthase gene. Mol Biochem Parasitol 83 (1):93-106.

[102] Mazumdar S, Sachdeva S, Chauhan VS, Yazdani SS. 2010. Identification of cultivation condition to produce correctly folded form of a malaria vaccine based on Plasmodium falciparum merozoite surface protein-1 in Escherichia coli. Bioprocess and biosystems engineering 33 (6):719-730. DOI: 10.1007/s00449-009-0394-x.

[103] Baca AM, Hol WG. 2000. Overcoming codon bias: a method for high-level overexpression of Plasmodium and other AT-rich parasite genes in Escherichia coli. Int J Parasitol 30 (2):113-118.

[104] Sano G, Morimatsu K, Horii T. 1994. Purification and characterisation of dihydrofolate reductase of Plasmodium falciparum expressed by a synthetic gene in Escherichia coli. Mol Biochem Parasitol 63 (2):265-273.

[105] Cinquin O, Christopherson RI, Menz RI. 2001. A hybrid plasmid for expression of toxic malarial proteins in Escherichia coli. Mol Biochem Parasitol 117 (2):245-247.

[106] Lindenthal C, Klinkert MQ. 2002. Identification and biochemical characterisation of a protein phosphatase 5 homologue from Plasmodium falciparum. Mol Biochem Parasitol 120 (2):257-268.

[107] Hoffman SL, Rogers WO, Carucci DJ, Venter JC. 1998. From genomics to vaccines: malaria as a model

system. Nat Med 4 (12):1351-1353. DOI: 10.1038/3934.

[108] Ryan W, Parulekar SJ. 1991. Recombinant protein synthesis and plasmid instability in continuous cultures of Escherichia coli JM103 harboring a high copy number plasmid. Biotechnol Bioeng 37 (5):415-429. DOI: 10.1002/bit.260370504.

[109] Hellmuth K, Korz DJ, Sanders EA, Deckwer WD. 1994. Effect of growth rate on stability and gene expression of recombinant plasmids during continuous and high cell density cultivation of Escherichia coli TG1. J Biotechnol 32 (3):289-298.

[110] Gupta R, Sharma P, Vyas VV. 1995. Effect of growth environment on the stability of a recombinant shuttle plasmid, pCPPS-31, in Escherichia coli. J Biotechnol 41 (1):29-37.

Chapter 3

Prospects of Biocatalyst Purification Enroute Fermentation Processes

Michael Bamitale Osho and Sarafadeen Olateju Kareem

Abstract

Biotransformation of broth through fermentation process suffers a major setback when it comes to disintegration of organic substrates by microbial agents for industrial applications. These biocatalysts are in crude/dilute form hence needs to be purified to remove colloidal particles and enzymatic impurities thus enhancing maximum activity. Several contractual procedures of concentrating dilute enzymes and proteins had been reported. Such inorganic materials include ammonium sulphate precipitation; salting, synthetic polyacrylic acid; carboxymethyl cellulose, tannic acid, edible gum and some organic solvents as precipitants etc. The emergence of organic absorbents such as sodom apple (*Calostropis procera*) extract, activated charcoal and imarsil had resulted in making significant impact in industrial circle. Various concentrations of these organic extracts have been used as purifying agents on different types of enzyme vis: lipase, amylase, protease, cellulase etc. Purification fold and stability of the enzyme crude form attained unprecedented results.

Keywords: Purification, Enzyme, Stability, *Calostropis procera* extract, Activated charcoal, Imarsil

1. Introduction

The fermentation process involves actual growth of the microorganism and product formation under agitation and aeration, optimum environmental conditions to provide uniform and adequate oxygen to the cell for growth and survival. A fermentation process is a biochemical process and, therefore, has requirements of sterility and use of cellular enzymatic reactions instead of chemical reactions aided by inorganic catalysts, sometimes operating at elevated temperature and pressure. Indigenous fermentation process draws the type attention of food scientists for taking the strategies of food security [1]. Microorganisms in fermented foods play major roles in health sector such as production of antimicrobial compounds, antioxidant, and probiotics properties.

The significance of fermentation of different substrates has gained attention in the beginning of the nineteenth century with profound resultant effects. Some products were produced via fermentation viz.; acetone, glycerol, lactic acid, butanol and baker's yeasts. Due to urgent need to treat the World War II army fighters, several metabolites such as amino acids, antibiotics and vitamins were produced via fermentation. Submerged fermentation with larger volumes under aerobic conditions with moderate process control was established during this period. In subsequent years, the fermentation industry has seen constant improvement with leaps and bounds on the production of high-value metabolites, including various antibiotics and growth hormones, using sophisticated bioreactors [2, 3].

Purification of enzyme is often a multifarious process and a number of procedures are normally employed in succession to obtain adequately high purity state. The idea is to use less expensive but simple methods at nascent stages when the volume is large and more expensive and advanced techniques when the volume is relatively small [4]. The aims are to obtain high final degree of purity; enhanced enzyme activity and reproducibility of the products. It obvious that extraction procedures release a number of other cell components like other enzymes, proteins, polysaccharides and nucleic acids apart from the target enzyme into the medium, this often resulted into increasing the viscosity of the solution depending on their polymeric structure. It is of great significance as to receive knowledge about functional and structural properties of the substance and to predict its applications.

2. Enzyme biosynthesis

Biocatalysis can be defined as the utilization of living materials or molecules to speed up the rate of chemical reactions. The usage of recombinant enzymes, groups of enzymes, naturally occurring enzymes, cell extracts and whole cells, modified or engineered enzymes inclusive. These biological materials have consequential edge over conventional chemical catalysis. Ideally, biochemical reactions occur in aqueous solution, at moderate temperatures and atmospheric pressure which can result in both environmental and economic values as compared with the existing processes at upraised temperatures and pressures, and in organic medium [5]. High cost of enzymes and potential environmental damages resulting from high temperatures and pressures requires substantial energy inputs. Comparatively little known aspect of biocatalysis is the role of protein potency compare to the overall average structure of a protein, mobility/potency is much difficult to estimate experimentally. Like other molecules, proteins are in sustained motion, with stretching, and rotating bonds bending. These motions can bring about to much greater, for example conformational rearrangements following substrate binding or the movement of two domains relative to one another. It is now clear that these motions play important in catalysis and the regulation of enzyme activity [6]. Moreover, recognizing those motions which come up with catalysis will be experimentally challenging. Likewise, modeling them precisely enough to predict their effects in novel arrangement will be tasking.

As important hydrolytic enzymes, amylase and protease represent the two largest groups of industrial enzymes and account for approximately 85% of total enzyme sales all over the globe. At present, more than 3000 different enzymes have been characterized and many of them found their way into biotechnological and industrial applications [7]. One technicality in enzyme technology especially those of starch biosynthetic enzymes is their manipulations to meet the enormous demand of teeming population, safeguard our environment from non-degradable biopolymer and of course satisfy the food industries' need [8]. Some commercially available biocatalysts do not resist industrial processing conditions due to severity of such conditions. Therefore, certain desirable characteristics during isolation and screening of novel enzymes viz.; alkaline stability, halophilicity, and thermostability are foremost to meet the industrial demand. Great deal of attention has been drawn on extremophiles, which are the valuable source of novel biocatalysts [9]. Extracellular enzymes from these halophiles with polymer-degrading ability at low water activity are of significant in many task processes where concentrated salt solutions hinder enzymatic production. The potency of enzymes to maintain its activeness via organic solvents has attracted considerable interest over the past twenty years.

In contrast to in water, numerous advantages of using enzymes in organic solvents or aqueous solutions containing organic solvents have been observed. Generally, enzymes are easily denatured and their activities disappear in the presence of organic solvents. Therefore, enzymes that remain stable in the presence of organic solvents might be useful for biotechnological applications in which such solvents are used [10]. Because salt reduces water activity, a feature in common with organic solvent systems, halophilic enzymes are thought to be valuable tools as biocatalysts in other low-water-activity environments, such as in aqueous/organic and non-aqueous media.

2.1 Purification protocols and applications

There are considerable ways of maintaining enzyme efficiency. Purification technique is a very powerful device has been arduously used to make some economically attainable and high performance enzymes with improved stability [11]. Enzyme purification is imperative for a full apprehension of the description and established process of enzymes. This is usually a multistep process involving biomass separation, concentration, primary isolation, and purification [12]. The contractual methods for the removal of enzymatic debrises or impurities and colloidal particles from fermentation broth include ammonium sulphate precipitation which encompassing dialysis for almost 16 hours before product could be recovered and also results into protein denaturation due to conformational changes [13].

Large volume of industrial enzymes are usually not purified. Their recovery is often accomplished by an ultrafiltration step. During enzyme production, desired products are synthesis after several concentration and separation techniques known as downstream processing (DSP). Two factors (time and cost) are the major challenges confronting these conventional techniques and their sustainability and efficiency depend on precise choice of purification methods [14, 15]. Here are some examples of strategies undertaken to improve the performance of enzymes with applications in food industry. Wong et al. [16] investigated strategies employed in starch liquefaction with targeted improvement of thermostability using α-amylase, protein engineering through site-directed mutagenesis and mutant displayed increased half-life between 15 min and 70 min at 100°C evolved. Glucoamylase with specific role as starch saccharifier and targeted improvement of substrate specificity, thermostability and pH optimum was characterized with protein engineering through site-directed mutagenesis alone [17]. Xylose (glucose) isomerase displayed isomerization/epimerization of hexoses, pentoses and tetroses as significance role of pH-activity profile with targeted improvement which resulted in protein engineering through directed evolution and the yield number on D-glucose in wild type was sustained between pH 6.0 and 7.5 and improvably at pH 7.3 as compared with mutant strains enhanced by 30–40%.

The application of polyvinyl alcohol or carbowax for protein and enzyme concentration is being restricted by poor water holding capacity. Moreover, gel filtration technique is also considered arduous and costly to the developing nations [18]. Carboxy-methyl cellulose, tannic acid, edible gum and some organic solvents as precipitants also poses the problem of product recovery [13]. In fishing industry, the use of fast, simple and low cost techniques such as using organic solvents vis ethanol and acetone, successive stages of centrifugation and filtration; and saline solution (ammonium sulfate) were adopted for the separation and partial purification of protein biomolecules obtained from fish by-products beneficiation remains [19–21], aiming to improve the degree of biomolecule purity [21]. The use of ammonium sulphate precipitates during enzyme purification needs protracted separation technique between 12 and 16 h for recovery of product that frequently bring about protein denaturation as gel chromatography is high priced and moderate for developing economies [22]. Application of chromatographic techniques such as gel filtration and ion exchange give rise to purer enzyme fractions, with significant increase in specific activity. These are often used to estimate the molecular mass of the enzyme by comparing protein mixtures of known molecular mass (reference standards) with the unknown.

Pectinase enzyme was precipitated by dissolving it in a 0.1 M, pH 4.2 sodium acetate buffer after mixing with 3 volumes of ice-cold acetone and allowed to stay for 15 min [23]. Chimbekujwo et al. [24] reported the application of SDS-PAGE analysis of purified fungal protease of major protein band with molecular weight of 68 KDa, 13.3 fold and 28% yield. This partially purified enzyme was stable between 30 and 40°C temperature and pH 4–6 which enhanced the activity by Tween-20 and Calcium ions. Moreover, during the production, characterization and anti-cancer application of extracellular L-glutaminase from the marine bacterial isolate, the enzyme was purified through QFF technique by engaging ethanol precipitation and ion-exchange chromatography and resulted into 2-fold purification with molecular weight 54.8 kDa, specific activity 89.78 U/mg, maximum enzyme activity at 40°C and pH 8.2 and ultimately retained 90% activity for an hour [25].

3. Adoption of organic absorbent materials for purification procedures

The emergence of organic materials such as activated charcoal, *Calostropis procera* latex and imarsil has made significant contributions for industrial applications. Some of these organic absorbents are discussed below:

3.1 Activated charcoal and other carbon particles as purifying agents

Activated charcoal is an adsorbent extensively utilized in the treatment of wastewater and industrial contaminants by reason of its high shifting ability and adaptability for a wide range of pollutants. It is produced from any crucially carbonaceous materials. Coal, cotton waste, tree barks, palm kernel shell, and many agricultural by- products can be made to produce activated carbon and their capacity to remove colors has been investigated. Ferreira and coworkers [26] demonstrated the production, characterization of activated charcoal from castor seed cake through activation with phosphoric acid. Treatment of fino sherry wine with activated charcoal, in combination with other clarifying agents, produces a wine with lower polyphenolic content, good organoleptic characteristics, but its receptive to browning is indistinguishable to that recognized in untreated wine, despite starting from lower levels of color potency. Activated charcoal is used to remove compounds that cause objectionable color, odor and taste in water treatment while its industrial applications require elimination of harmful gases and pesticides and including purification of organic compounds [27]. It is established that 80% of activated charcoal globally produced is used in aqueous-phase adsorption of both organic and inorganic compounds [28]. However, the application of activated charcoal in the decolorization of enzyme-converted glucose syrup had been described; though its application for the purification of microbial biocatalysts has been sparse.

One profound advantage of activated charcoal over conventional purification systems is this swift enzyme purification from composite fermentation broth mixture at a very high purification fold. These conventional procedures of purification of enzyme among others include solvent precipitation; gel filtration and salting out technique. From an industrial application stand point of view, they are quite expensive base on the fact that they are associated with some difficulty of scaling up and plugging leading to viscous and particulate materials when treated with crude enzyme extracts. Additionally, it is not economical for developing nations as the materials disposal or enzyme recovery techniques employed in the separation method might escalate the expenses. As a result, the usage of activated charcoal has been considered as preferred option for enzyme purification method.

In addition to the inexpensiveness of activated charcoal, their efficient surface absorption attributes can be exploited for depolarization of fermented medium for efficacious and efficient recovery and purification of industrial enzymes making the downstream processing in large-scale industrial bioprocesses less economical [29]. López et al. [30] in their investigation on the use of activated charcoal in combination with other fining agents as clarifying agents reported that these carbonized materials acted upon the phenolic compounds thus encouraging their precipitation. In the field of enology, many different substances have been employed as fining agents over time such addition of antioxidants (ascorbic acid, sulfur dioxide and bottling under inert atmosphere. The use of bentonite has been well-proven and reported to have a remarkable effect on the protein content of wine and also hastens the precipitation of the thermolabile protein [31] but has also minimized the polyphenolic content of the wine during production [32].

The structure of activated carbon which is based on the graphite lattice corresponds to a non-graphitizable carbon and macromolecular structure of the precursor residues during heat treatment, and losses small molecules by developing and degradation some cross-linking, so that joining cannot occur. Therefore, cross-linking bring about a fixed design with small vigor, thus producing a permeable system and intercepting the ordering expected during graphitization.

3.2 *Calotropis procera* as purifying agent and its industrial applications

Calotropis procera belongs to the family Asclepidaceae being a native of tropical and subtropical region of Africa, the Middle East, and South and South-East Asia [33]. It is a shrub that produces latex with wide pharmacological profile which is a rich source of biologically active compounds [34]. *C. procera* latex contain several chemical compounds which include calotropagenin glycosides/derivatives [35]; saponins, flavonoids and cardenolides [36, 37]; cardioids such as calotoxin, calotropin, uscherin, uscchardin, choline, o-pyrocatechuric acid, glycoside calotropaginin, benzoyllineolone, benzoylisoloneolane, syriogenis and uzariganin etc. [38]. It has been traditionally used for various medicinal purposes such as treatment of animal worms, defense role in plants, acting against herbivorous insects, nematodes and phytopathogenic fungi [39]. Different parts of roots, leaves, flowers and latex from the plant are used in several medicinal preparations [40]. It was also reported to exhibit potent analgesic and weak antipyretic activity in various experimental model, possess antioxidant and anti-hyperglycemic property [41], antihelmintic activity [42, 43]; insecticidal and antifungal proteins and their enzymatic profiles have been characterized [44–46]; and there is an empirical association between antioxidant property and residual peroxidase activity. The milk weed has been established to be efficacious in the chemotherapy of malaria, menorrhagia, fever, leprosy and snake bites. Research works investigated on many biological activities of *C. procera* including osmotin proteins exert antifungal activity [47] and anti-inflammatory potential in rats. *C. procera* latex dispensed to rats revealed pain-killing effects wound healing and toxic [48].

The leaf of *Calotropis procera* is a natural coagulant used traditionally in waste water treatment and it has also been reported that *Calotropis procera* leaf is effective in removal of environmental pollutant, polyphenolic crystal violet dye from aqueous solution of textile effluent [49] which presumed to be ascribed to the presence of peroxidase in *Calotropis procera* leaf that oxidized phenols to phenoxy radicals. Some studies suggest that the insoluble fraction of *C. procera* latex is associated with the noxious effects of this fluid [50]. Contrariwise, some constituents of this fluid cause toxicity in small ruminants [51].

From the investigation carried out by Mafulul *et al.*, [52] in the extraction, partial purification and characterization of peroxidase from *Calotropis procera* leaves, it was revealed that peroxidise purified from *Calotropis procera* leaves in primary purification procedures resulted in 1.613-fold purification of peroxidase from the crude extract. Subsequently, enzyme precipitation using ammonium sulphate with the dialyzed fraction showed 2.04 purification folds. *Calotropis procera* leaves peroxidase maintained above 50% over a temperature range of 20–70 with optimum temperature 50°C.

Furthermore, considering the availability and abundance of *Calotropis procera* fresh leaves in Nigerian distribution coupled with availability of advance purification method, this plant tends to provide a very cheap source of peroxidase for phenolic pollutants' bioremediation for waste treatment especially in oil spill region of Niger Delta. It provides potential alternative peroxidase that can compete with commercially available peroxidases for biotechnological applications.

3.3 Imarsil - an inexpensive synthetic chromatographic absorbent

Imarsil is a novel, inexpensive synthetic chromatographic absorbent and oxidized natural polymer of *Brachystegia nigerica. B. nigerica* is a legume used especially in the eastern states of Nigeria as condiment to thicken soup. Its thickening characteristics have been attributed to the presence of hydrocolloid property or gelling property [53]. Imarsil possesses quick and simple recovery approach more importantly in the clarification of microbial biocatalyst from fermentation broth [11]. Cherry and Fidants, [54] demonstrated the use of carboxy-methyl cellulose, edible gum and tannic acid as precipitants and as well as organic solvents also poses the problem of product recovery. Gel filtration technique is also considered assiduous and expensive in the developing countries.

Several procedures of concentrating protein and dilute enzyme from fermentation cell extracts and media using agricultural residues as coagulants. Furthermore, Kareem *et al.* [55] investigated the use of Imarsil and activated charcoal to purify crude lipase in a two-step purification fold which brought about an increase in specific activity from 5.29 to 20.8 Umg^{-1} with protein reduction of 18.24% in the supernatant and ultimate 3.93-fold purification. The study on crude amylase purification showed that a 40-fold purification was attained with 50% final yield of the total fungi amylase in a 3-step purification technique. The elution pattern of *Rhizopus oligosporus* SK5 amylase on Sephadex G-100 column had peaks at fractions (19–22) and (34–38). This purification fold value is conceived greater than values obtained in previous work [56].

Osho et al. [57] studied on production and optimization of bacterial cellulase using agricultural cellulosic biomass by solid state cultivation where the enzyme was clarified with Whatman No 1 filter paper, partially purified with Imarsil (1% w/v) and incubated at 4°C for 3 h. It was reported that at temperature ranges of 40–90°C, enzyme activity increases in crude and partially clarified states as the relative activity also increased to 50 and 60°C for both forms of cellulase respectively. A decline in activity was noticed as temperature increases for both solutions. However, 90% activity of the partially purified enzyme was retained between 50 and 55°C and activity peaked at 60°C. Partial clarification of enzyme is therefore needed to enhance their stability even at much temperature. Kareem and other coworkers [58] outlined that partial clarification of enzyme using activated charcoal preceding gel filtration will established a high purification fold thus preventing some awkwardness of plugging and scaling up when treating crude extracts that sometimes contain particulate and viscous materials. These studies have further substantiated the use of Imarsil as a coagulating-flocculating agent in purification of crude enzyme extracts.

4. Conclusion

It has been proven that enzymes could be recovered from the fermentation broth by these organic absorbents and flocculating materials, making fermentation procedure less laborious. Following elution process, a highly concentrated and purified enzyme would be obtained at reasonable time. This technique seems to be rapid, cheap and promising in downstream processing of industrial enzymes which leads to an aqueous enzyme concentration. They are also established to be faster and easier to implement than the two or three-step processes of conventional precipitation, dialysis and subsequent chromatography. Thus, these natural coagulating-flocculating materials are of great importance in that they are effective in removal of pollutants and debris from fermented broths without necessarily affects the functional and structural formation of industrial enzymes.

Conflict of interest

The authors declare no conflict of interest.

Author details

Michael Bamitale Osho[1]* and Sarafadeen Olateju Kareem[2]

1 Department of Biological Sciences (Microbiology Unit), College of Natural and Applied Sciences, McPherson University, Seriki Sotayo, Abeokuta, Ogun State, Nigeria

2 Department of Microbiology, Federal University of Agriculture, Abeokuta, Nigeria

*Address all correspondence to: mikebamosho@gmail.com

References

[1] van de Sande T. Socio-economic pitfalls of enhancing indigenous capabilities in household fermentation. Food Control 1997; 303-310. https://doi.org/10.1016/S0956-7135(97)00053-4

[2] Motarjemi Y, Nout MJ. Food fermentation: a safety and nutritional assessment. Joint FAO/WHO Workshop on Assessment of Fermentation as a Household Technology for Improving Food Safety. Bull World Health Organ 1996; 74 (6):553-59. https://apps.who.int/iris/handle/10665/54322

[3] Li S, Li P, Feng F, Luo LX. Microbial diversity and their roles in the vinegar fermentation process. Applied Microbiology and Biotechnology. 2015; 99 (12):4997-5024. http://dx.doi: 10.1007/s00253-015-6659-1

[4] Ramos OL, Malcata FX. Industrial Biotechnology and Commodity Products. In Comprehensive Biotechnology (Third Edition), 2017

[5] Timson DJ. Four Challenges for Better Biocatalysts Review. *Fermentation* 2019, 5, 39; http://dx.doi:10.3390/fermentation5020039

[6] Ahmed R, Saradar M. Immobilization of cellulase on TiO2 nanoparticles by physical and covalent Methods: A comparative study. International Journal of Biochemistry, Biophysics and Molecular Biology. 2014; 51: 314-20.

[7] van den Burg B. Extremophiles as a source for novel enzymes. Current Opinion Microbiology. 2003; 6: 213-218. https://dx.doi:10.1016/s1369-5274(03)00060-2

[8] Verma VC, Agrawal S, Kumar A, Jaiswal JP. Starch content and activities of starch biosynthetic enzymes in wheat, rice and millets. Journal of Pharmacognosy and Phytochemistry 2020; 9(4): 1211-1218 https://doi.org/10.22271/phyto.2020.v9.i4q.11883

[9] Antranikian G, Vorgias CE, Bertoldo C. Extreme environments as a resource for microorganisms and novel biocatalysts. Advanced Biochemistry and Engineering Biotechnology. 2005; 96: 219-262. https://dx.doi: 10.10.1007/b135786

[10] Shafiei P, Karimi K, Taherzadeh MI. Techno-economical study of ethanol and biogas from spruce wood by NMMO-pretreatment and rapid fermentation and digestion. Bioresource Technology. 2011; 102:7879-7886. https://dx.doi:10.1016/j.biortech.2011.05.071

[11] Kareem SO, Adio OQ, Osho MB. Immobilization of Aspergillus niger F7-02 Lipase in Polysaccharide Hydrogel Beads of Irvingia gabonensis Matrix. Enzyme Research Article ID 967056, 2014; 7 pp. http://dx.doi.org/10.1155/2014/967056

[12] Kareem SO, Akpan I, Osho MB. Calotropis procera (Sodom apple)—a potential material for enzyme purification. Bioresource Technology. 2003; 87(1):133-135. https://dx.doi:10.1016/S0906-8524(02)00208.0

[13] Banerjee S, Maiti TK, Roy RN. Production, purification, and characterization of cellulase from Acinetobacter junii GAC 16.2, a novel cellulolytic gut isolate of *Gryllotalpa africana*, and its effects on cotton fiber and sawdust. Annals of Microbiology 2020. 70:28 https://doi.org/10.1186/s13213-020-01569-6

[14] Wang SS, Ning YJ, Wang SN, Zhang J, Zhang GQ, Chen QJ. Purification, characterization, and cloning of an extracellular laccase with

potent dye decolorizing ability from white rot fungus *Cerrena unicolor* GSM-01. International Journal of Biological Macromolecules 2017. 95: 920-927

[15] Antecka A, Blatkiewicz M, Boruta T, Górak A,Ledakowicz S. Comparison of downstream processing methods in purification of highly active laccase. Bioprocess and Biosystems Engineering. 2019. 42:1635-1645. https://doi.org/10.1007/s00449-019-02160-3

[16] Wong DWS, Batt SB, Lee CC, Robertson GH. High-activity barley α-amylase by directed evolution. Protein Journal. 2004; 23(7): 453-460.

[17] Allen MJ, Fang TY, Li, Y. Protein engineering of glucoamylase to increase pH optimum, substrate specificity and thermostability. United States Patents No. 6 2003; 537,792.

[18] Kareem SO, Akpan I. Clarification of amylase extract from moldy bran with Imarsil. Enzyme and Microbial Technology 2003; 33(2-3): 259-261. https://dx.doi:10.1016/S0141-0229(03)00127-3

[19] Khawli FAl, Ferrer E, Berrada H, Barba FJ, Pateiro M, Domínguez R, Lorenzo JM, Gullon P, Kousoulaki K. Innovative green technologies of intensification for valorization of seafood and their by-products. Mar. Drugs 2019. 17; 2-21. https://doi.org/10.3390/md17120689

[20] Oliveira VD, da Cunha MNC, de Assis CRD, Batista JMD, Nascimento TP, dos Santos JF, Lima CD, Marques DDV, Bezerra RD, Porto ALF. Separation and partial purification of collagenolytic protease from peacock bass (*Cichla ocellaris*) using different protocol: Precipitation and partitioning approaches. Biocatalysis and Agricultural Biotechnology 2020. 24. https://doi.org/10.1016/j.bcab.2020.101509

[21] Oliveira VM, Assis CRD, Silva JC, Silva QJ, Bezerra RS, Porto ALF. Recovery of fibrinolytic and collagenolytic enzymes from fish and shrimp byproducts: potential source for biomedical applications. Bol. Inst. Pesca 2019. 45; 389. https://doi.org/10.20950/1678-2305.2019.45.1.389

[22] Kareem SO, Akpan I. Clarification of amylase extract from moldy bran with Imarsil. Enzyme and Microbial Technology 2003; 33(2-3): 259-261. https://dx.doi:10.1016/S0141-0229(03)00127-3

[23] Sudeep KC, Upadhyaya J, Joshi DJ, Lekhak B, Chaudhary DK, Pant BR, Bajgai TR, Dhital R, Khanal S, Koirala N, Raghavan V. Production, Characterization, and Industrial Application of Pectinase Enzyme Isolated from Fungal Strains. Fermentation 2020. 6 (59); http://dx.doi:10.3390/fermentation6020059

[24] Chimbekujwo KI, Ja'afaru MI, Adeyemo OM. Purification, characterization and optimization conditions of protease produced by *Aspergillus brasiliensis* strain BCW2 Journal of Scientific African 2020. 8; e00398 https://doi/org/10.1016/j.sciaf.2020.e00398

[25] Orabi H, El-Fakharany E, Abdelkhalek E, Sidkey N. Production, optimization, purification, characterization, and anti-cancer application of extracellular L-glutaminase produced from the marine bacterial isolate. Prep Biochemistry and Biotechnology 2020. 50(4); 408-418. https://dx.doi:10.1080/10826068.2019.1703193

[26] Ferreira L.M, de Melo RR, Pimenta AS, de Azevedo TKB, de Souza CB. Adsorption performance of

activated charcoal from castor seed cake prepared by chemical activation with phosphoric acid. *Biomass Conversion Biorefinery* 2020. https://doi.org/10.1007/s13399-020-00660-x

[27] Ochonogor AE. Indigo carmine removal from solutions using activated carbons of Terminalia catapa and Cinnarium schweinfurthi nutshell. Journal of Chemical Society of Nigeria. 2005; 30(2): 88-90

[28] Hayashi J, Yamamoto N, Horikawa T, Muroyama K, Gomes VG. Preparation and characterization of highspecific-surface-area activated carbons from K2CO3-treated waste polyurethane. Journal of Colloid and Interface Science. 2005; 281(2): 437-443.

[29] Kumar CG, Takagi H. Microbial alkaline proteases: from a bioindustrial viewpoint. Biotechnology Advances. 1999; 17(7): 561-594. https://dx.doi:10.1016/S0734-9750(99)00027-0

[30] López S, Castro R, García E, Pazo JSA, Barroso CG. The use of activated charcoal in combination with other fining agents and its influence on the organoleptic properties of sherry wine. European Food Research and Technology. 2001; 212:671-675 http://dx.doi:10.1007/s002170100300

[31] Al Sulaibi MAM, ThiemannC, Thiemann T. Chemical Constituents and Uses of *Calotropis procera* and *Calotropis gigantea* – A Review (Part I – The Plants as Material and Energy Resources) Open Chemistry Journal. 2020; *7*, 1-15 http://dx.doi:10.2174/1874842202007010001

[32] Main GL, Morris JR. Colour of seyval blanc juice and wine as affected by juice fining and bentonite fining during fermentation. American Journal of Enology and Viticulture. 1994; 45:417-422.

[33] Al Sulaibi MAM, ThiemannC, Thiemann T. Chemical Constituents and Uses of *Calotropis procera* and *Calotropis gigantea* – A Review (Part I – The Plants as Material and Energy Resources) Open Chemistry Journal. 2020; *7*, 1-15 http://dx.doi:10.2174/1874842202007010001

[34] Kumar VL, Arya S. Medicinal uses and pharmacological properties of *Calotropis procera*. In: Recent Progress in Medicinal Plants, 11. (Govil E. D., ed.). Studium Press, Houston, TX, USA. 2006; 373-388 pp.

[35] Chundattu SJ, Agrawal VK, Ganesh N. Phytochemical investigation of *Calotropis procera*. Arabian Journal of Chemistry. 2011; https://doi.org/10.1016/j.arabjc.2011.03.011

[36] Yoganandam K, Ganeshan P, NagarajaGanesh B, Raja K. Characterization studies on *Calotropis procera* fibers and their performance as reinforcements in epoxy matrix. Journal of Natural Fibers. 2019. http://dx.doi.org/10.1080/15440478.2019.1588831

[37] Patil **SP.** Calotropis gigantea *assisted green synthesis of nanomaterials and their applications: a review.* Beni-Suef University Journal of Basic and Applied Sciences 2020. 9:14 https://doi.org/10.1186/s43088-020-0036-6

[38] Tu LX, Duan WZ, Xiao WL, Fu CX, Wang AQ, Zheng Y. *Calotropis gigantea* fiber derived carbon fiber enable fast and efficient absorption of oils and organic solvents. Separation Purification Technique 2018. *192*; 30-35. http://dx.doi.org/10.1016/j.seppur.2017.10.005

[39] Xiao, W.; Wang, N.; Niu, B.; Fu, C.; Zhou, L.; Zheng, Y. Polyethylene sulfone assisted shape construction of *Calotropis gigantea* fiber for preparing a sustainable and reusable oil sorbent. Cellulose, 2019. *26;* 3923-3933. http://dx.doi.org/10.1007/s10570-019-02356-6

[40] Cao, E.; Duan, W.; Yi, L.; Wang, A.; Zheng, Y. Poly(mphenylenediamine) functionalized *Calotropis gigantea*n an

fiber for coupled adsorption reduction for Cr(VI). Journal of Molecular Liquids 2017. *240*; 225-232. http://dx.doi.org/10.1016/j.molliq.2017.05.087

[41] Cao E, Duan W, Wang A, Zheng Y. Oriented growth of poly (mphenylenediamine) on *Calotropis gigantea* fiber for rapid adsorption of ciprofloxacin. Chemosphere 2017. *171*; 223-230. http://dx.doi.org/10.1016/j.chemosphere.2016.12.087

[42] Yi LS, Liang GW, Xiao WL, Duan WZ, Wang AQ, Zheng Y. Rapid nitrogen-rich modification of *Calotropis gigantea* fiber for highly efficient removal fluoroquinolone antibiotics. Journal of Molecular Liquids. 2018. *256*; 408-415. http://dx.doi.org/10.1016/j.molliq.2018.02.060

[43] Kaur R, Kaur H. *Calotropis procera* an effective adsorbent for removal of Congo red dye: isotherm and kinetc modelling. Model. Earth System Environment 2017. *3* http://dx.doi.org/10.1007/s40808-017-0274-3

[44] Alshahrani NDSTS, Aref IM, Nasser RA. Allelopathic potential of *Calotropis procera* and Eucalyptus species on germination and growth of some timber trees. *Allelopathy Journal* **2017**. *40*; 81-94. http://dx.doi.org/10.26651/2017-40-1-1068

[45] Cavalcante GS, Morais SM, Andre WPP, Ribeiro WLC, Rodrigues ALM, Lira FCML. Chemical composition and in vitro activity of *Calotropis procera* (Ait.) latex on Haemonchus contortus. Veterinary Parasitology. 2016; 226:22-25. http://dx.doi.org/10.1016/j.vetpar.2016.06.012

[46] Cavalcante GS, Morais SM, André WPP, Araújo-Filho JV, Muniz CR, Rocha LO. Chemical constituents of *Calotropis procera* latex and ultrastructural effects on Haemonchus contortus. Brazilian Journal of Veterinary Parasitology. 2020; 29(2): e001320. https://doi.org/10.1590/S1984-29612020045

[47] Doshi H, Satodiya H, Thakur MC, Parabia F, Khan A. Phytochemical screening and biological activity of *Calotropis procera* (Ait). R.Br. (Asclepiadaceae) against selected bacteria and Anopheles stephansi Larvae. International Journal of Plant Research. 2011; 1(1): 29-33. https://doi:10.5923/j.plant.20110101.05

[48] Shaker KH, Morsy N, Zinecker H, Imhoff JF, Schneider B. Secondary metabolites from *Calotropis procera* (Aiton). Phytochemistry Letters. 2010; 3(4): 212-216.

[49] Ali H, Muhammad SK. Biosorption of crystal violet from water on leafe biomass of *Calotropis procera*. Journal of Environmental Science and Technology. 2008; 1(3):143-150. http://dx.doi.org/10.3923/jest.2008.143.150

[50] Kumar VL, Shivkar YM. In vivo and in vitro effect of latex of *Calotropis procera* on gastrointestinal smooth muscles. Journal of Ethnopharmacology. 2004; 93(2-3): 377-379. http://dx.doi.org/10.1016/j.jep.2004.04.013

[51] El Sheikh HA, Ali BH, Homeida AM, Hassan T, Idris OF, Hapke HJ. The activities of drug-metabolizing enzymes in goats treated orally with the latex of *Calotropis procera* and the influence of dieldrin pretreatment. Journal of Comp Pathology. 1991; 104(3): 257-268. http://dx.doi.org/10.1016/S0021-9975(08)80038-2

[52] Mafulul SG, Joel EB, Barde LA, Lenka JL, Ameh AA, Phililus MG. Extraction, Partial Purification and Characterization of Peroxidase from *Calotropis procera* Leaves. Journal of Advances in Biology & Biotechnology. 2018; 18(1): 1-10. http://dx.doi:10.9734/JABB/2018/41709

[53] Odum DC. Proximate composition and Sugars of the hydrocolloid extract of Brachystegia nigerica. Nigerian Journal of Microbiology. 2000; 14:77-82.

[54] Cherry JR, Fidants AL. Directed evolution of industrial enzymes: An update. Current Opinion Biotechnology. 2003; 4: 438-443. http://dx.doi.org/10.1016/s0958-1669(03)00099-5

[55] Kareem SO, Akpan I, Popoola TOS, Sanni LO. Purification and characterization of thermostable glucoamylase from Rhizopus oligosporus SK5 mutant obtained through UV radiation and chemical mutagenesis. Biokemistri. 2014; 26 (1) 19-24.

[56] Selvakumar PA, Shakumary L, Helen A, Pandey A. Purification and Characterization of glucoamylase produced by Aspergillus niger in solid state fermentation. Letters in Applied Microbiology. 1996; 23: 403-406. http://dx.doi.org/10.1111/j.1472-765x

[57] Osho MB, Nwagala PN, Ojo EE. Production and Optimization of Bacterial Cellulase Using Agricultural Cellulosic Biomass by Solid State Cultivation. Current Biotechnology. 2017; 6: 349-355. http://dx.doi.org/10.2174/2211550106666170109151156

[58] Kareem SO, Adebayo OS, Balogun SA, Adeogun AI, Akinde SB. Purification and Characterization of Lipase from Aspergillus flavus PW2961 using Magnetic Nanoparticles. Nigerian Journal of Biotechnology. 2017; 32:77-82 http://dx.doi.org/10.4314/njb.v32i1.11.

Chapter 4

Integrated Biorefinery Approach to Lignocellulosic and Algal Biomass Fermentation Processes

Felix Offei

Abstract

Lignocellulosic and algal biomass have been suggested as relatively sustainable alternatives to sugar and starch-based biomass for various fermentation technologies. However, challenges in pretreatment, high production costs and high waste generation remains a drawback to their commercial application. Processing cellulosic and algal biomass using the biorefinery approach has been recommended as an efficient and cost-effective pathway since it involves the recovery of several products from a single biomass using sequential or simultaneous processes. This review explored the developments, prospects and perspectives on the use of this pathway to add more value and increase the techno-economic viability of cellulosic and algal fermentation processes. The composition of lignocellulosic and algal biomass, the conventional ethanol production processes and their related sustainability issues are also discussed in this chapter. Developments in this approach to lignocellulosic and algal biomass has shown that valuable products at high recovery efficiencies can be obtained. Products such as ethanol, xylitol, lipids, organic acids, chitin, hydrogen and various polymers can be recovered from lignocellulosic biomass while ethanol, biogas, biodiesel, hydrocolloids, hydrogen and carotenoids can be recovered from algae. Product recovery efficiencies and biomass utilisation have been so high that zero waste is nearly attainable. These developments indicate that indeed the application of fermentation technologies to cellulosic and algal biomass have tremendous commercial value when used in the integrated biorefinery approach.

Keywords: fermentation, integrated biorefinery, lignocellulosic biomass, algae, ethanol

1. Introduction

Concerns over the depletion and environmental effects of greenhouse gas (GHG) emissions from the use of fossil fuels has led to the extensive search for alternative, renewable and sustainable fuels. Currently, the highest contributor to GHG emissions is the transportation sector through fuel combustion. Biomass is currently the only abundant renewable energy source for the direct production of fuel. Typical fuels currently produced from biomass include bioethanol, biogas, biodiesel, bio-butanol, syngas and bio-oil. Bioethanol is currently the largest alternative fuel produced globally at 106 billion litres per annum [1].

Sugar and starch-based biomass have been the primary choice of raw material for the production of food and fuel grade ethanol for various commercial applications. They however face enormous competing interests often illustrated with the food-vs.-fuel debate [2]. Lignocellulosic and algal biomass have been suggested as relatively sustainable alternatives. They have been hundreds of extensive research on the factors that influence their efficiency as substrates for ethanol production. The major drawbacks noted in these studies during their application include: the need for pretreatment processes, higher production costs and high waste generation [3]. A processing approach that has potential to maximise the profitability and minimise waste generation from the use of cellulosic and algal biomass as feedstock is the integrated biorefinery approach. The integrated biorefinery concept refers to the use of single or multiple technologies to produce several high value products from a single or multiple biomass [4]. This approach to biomass processing is considered more efficient, economical and sustainable.

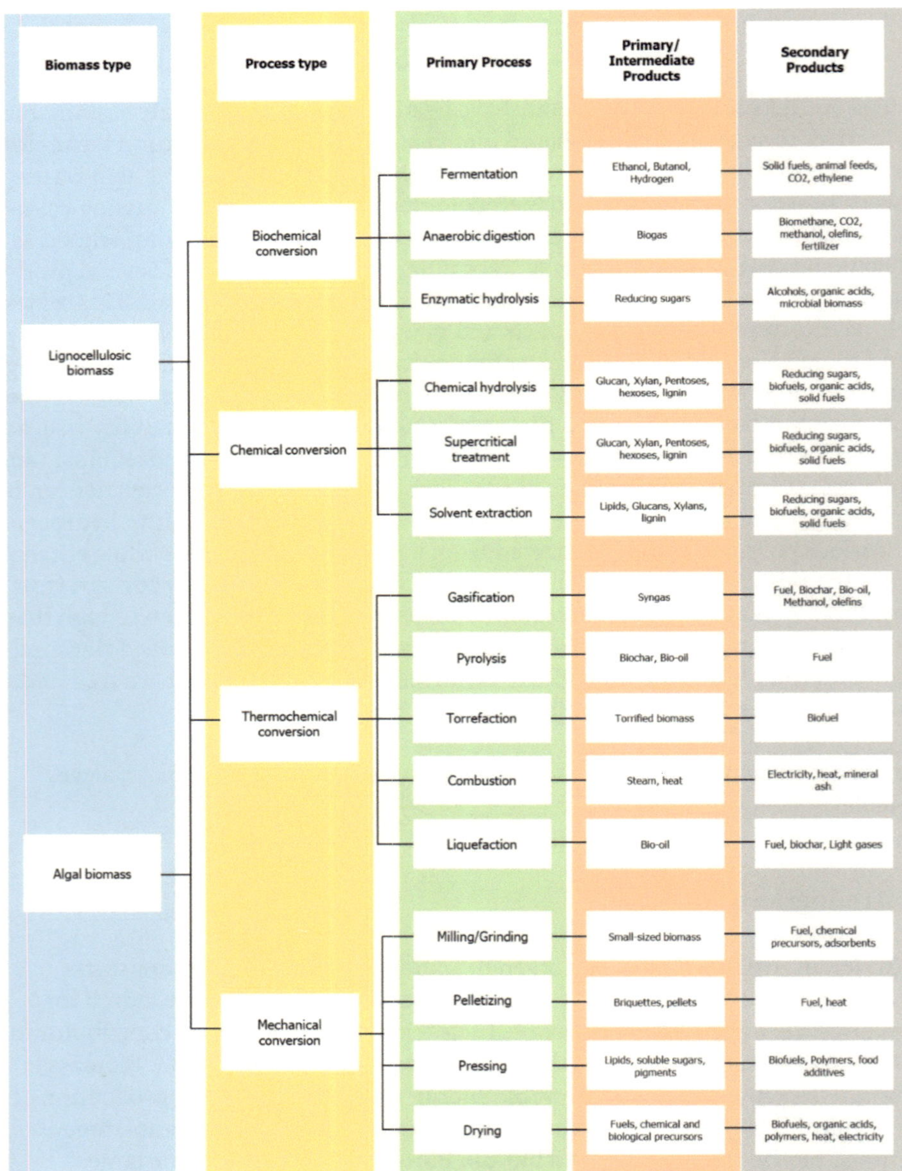

Figure 1.
Typical biorefinery conceptual scheme.

Biorefineries generally integrate various biomass conversion technologies to produce fuels, power, heat and other value-added products from biomass. These refineries have evolved over the last two decades in several phases. Phase I biorefineries convert a single raw material to a single product. Phase II converts a single raw material using multiple processing tools to obtain a broad range of products. Phase III biorefineries, commonly referred to as integrated biorefineries use a wide range of raw materials and technologies simultaneously or sequentially to produce a wide range of valuable products [5]. Some integrated biorefineries use various feedstock and technologies to produce biofuels as main products along with co-products such as platform chemicals, heat and power [5].

The International Energy Agency sums up the description of the biorefinery concept as "the sustainable processing of biomass into a spectrum of marketable products and energy" [6]. It expands the concept to include a wide range of technologies that separate biomass resources into their basic polymeric units such carbohydrates, proteins, lipids and even elementals which can be converted to valuable products including fuels, heat and chemicals. Biorefinery as an entity is described as a facility or network of facilities where various processing technologies are integrated to obtain multiple products from a single or several types of biomass [6]. Bioethanol is currently the leading energy product recovered from biomass using the biorefinery approach.

Sugar and starch-based biomass have been the primary choice of material for the production of food and fuel grade ethanol for various commercial applications but has an enormous competing interest often illustrated with the food-vs.-fuel debate. Lignocellulosic and algal biomass have been suggested as relatively sustainable alternatives. However, difficulties in pretreatment, high waste generation and high processing costs remains a drawback to their commercial application. Processing cellulosic and algal biomass using the biorefinery approach has been recommended as an efficient and cost-effective pathway since several valuable products can be recovered using sequential or simultaneous processes as illustrated in **Figure 1** [4]. This review explored the developments made in the use of this pathway to add more value and increase the techno-economic viability of cellulosic and algal fermentation processes. The composition of lignocellulosic and algal biomass, the conventional ethanol production processes and their related sustainability issues are also discussed in this chapter.

2. Lignocellulosic biomass for biorefinery applications

Lignocellulosic biomass typically refers to plant materials composed primarily of cellulose, hemicellulose and lignin. This type of biomass usually includes forest materials, agricultural residues, wood processing residues and non-edible plant materials usually referred to as energy crops (**Table 1**). In the context of biofuel production, lignocellulosic biomass are referred to as second generation biomass which is used to differentiate them from sugar and starch based biomass (1st generation biomass) and algal biomass (3rd generation biomass). They are typically composed of 40–50% cellulose, 25–30% hemicellulose and 15–20% lignin [30]. The effective use of these three primary components would significantly determine the economic viability of cellulosic ethanol production.

Cellulose refers to the linear polymer made up of glucose monomer units bonded together by β-1,4 glycosidic bonds. Hemicellulose refers to branched heteropolymers of xylose, glucose, galactose, mannose, arabinose and some uronic acids. Lignin is primarily made up of three major phenolic components, namely p-coumaryl alcohol, coniferyl alcohol and sinapyl alcohol [20]. The ratio of these

Biomass type	Biomass	Cellulose (%)	Hemicellulose (%)	Lignin (%)	Reference
Agricultural residues	Sugarcane bagasse	49	29.6	27.2	[7]
	Barley straw	37.5	37.1	16.9	[8]
	Rice husk	33.4	30.0	18.3	[9]
	Corn cob	44	36.4	18.0	[10]
	Corn stover	36.5	31.3	13.6	[11]
	Rye straw	42.1	24.4	22.9	[12]
	Rapeseed straw	37.0	24.2	18.0	[13]
	Wheat straw	40.0	33.8	26.8	[14]
	Rice straw	36.6	22.0	14.9	[15]
	Sunflower stalk	33.8	24.3	19.9	[16]
	Sorghum bagasse	45.3	26.3	16.5	[17]
	Barley hull	34.0	36.0	19.0	[18]
	Banana peels	13.0	15.0	14.0	[19]
	Cotton stalk	31.0	11.0	30.0	[20]
	Coffee pulp	36.9	47.5	19.1	[20]
	Wheat bran	14.8	39.2	12.5	[20]
	Sugarcane tops	35.0	32.0	14.0	[21]
	Jute fibres	45.0	18.0	21.0	[20]
	Oat straw	31.0	20.0	10.0	[20]
	Soya stalks	34.5	24.8	9.8	[22]
Municipal and industrial wastes	Newspapers	60.3	16.4	12.4	[23]
	Paper sludge	60.8	14.2	8.4	[24]
	Brewer's spent grain	21.0	32.8	25.6	[25]
Woods and grasses	Softwood stems	44.5	21.9	27.7	[6]
	Switchgrass	35.4	26.5	18.2	[6]
	Bamboo	50.0	20.0	23.0	[20]
	Eucalyptus	51.0	18.0	29.0	[20]
	Hardwood stems	55.0	40.0	25.0	[26]
	Pine	49.0	13.0	23.0	[20]
	Poplar wood	51.0	25.0	10.0	[20]
	Olive tree	25.2	15.8	19.1	[27]
	Water hyacinth	22.1	50.1	5.4	[28]
	Spruce	43.8	20.8	28.3	[29]
	Oak	45.2	24.5	21.0	[29]

Table 1.
Composition of typical lignocellulosic biomass used in biorefinery applications.

components varies between various plant tissues as shown in **Table 1**. The cellulose units are packed into microfibrils which are attached to each other by hemicelluloses and amorphous polymers of different sugars as well as other polymers such as pectin covered by lignin. The units of individual microfibrils in crystalline cellulose are packed so tightly that neither enzymes nor water molecules can enter the complex framework [20]. This high molecular weight and ordered tertiary structure of natural cellulose makes it insoluble in water. However, some parts of the microfibrils have a less ordered, non-crystalline structure referred to as amorphous regions [31]. The crystalline regions of cellulose are more resistant to biodegradation than the amorphous parts while cellulose with low degree of polymerisation will be more susceptible to cellulolytic enzymes. The composition of typical lignocellulosic biomass that have been considered for various biorefinery applications are presented in **Table 1**.

3. Algal biomass for biorefinery applications

Marine biomass accounts for over 50% of primary biomass produced globally but has been the least harnessed for various applications [32]. It is mainly grouped into two, namely macroalgae (commonly known as seaweeds) and microalgae. However, cyanobacteria is conventionally regarded as a form of algae often called blue-green algae [33]. Both groups have been used in the production of various biofuels. Microalgae has been explored predominantly as substrate for bio-oils and biodiesel while macroalgae has been used mainly in bioethanol and biogas production [32].

Marine algae are plant-like multicellular organisms that live attached to hard substrata such as rocks in coastal areas [34]. Their basic structure consists of a thallus, which forms the body of the organism and a holdfast, a structure on its base which allows it to be attached to hard surfaces such as rocks near the shoreline of coastal areas. Brown seaweeds are the largest in size, growing up to 4 m in length for some species. Green and red seaweeds are smaller ranging from a few centimetres in some species to a meter in others [35]. According to the FAO [36], 8.2 and 15.8 million tons of brown and red seaweed respectively were produced in the year 2013. This was valued at USD 1.3 billion and 4.1 billion for the brown and red seaweeds respectively. For the green seaweed 14,800 tons valued at USD 15.7 million was produced globally in the year 2013 [36]. The enormous difference in the production values of the brown and red from the green seaweed can be attributed to the valuable hydrocolloids such as alginate, carrageenan and agar found only in the red and brown seaweeds.

The structural differences found between land-based plants and algae gives algal biomass an advantage of a higher yield per hectare. In comparison to land-based plants, seaweeds have an average yield per hectare per year of 730,000 kg while sugarcane, sugar beet, maize and wheat have 68,260; 47,070; 4,815 and 2,800 kg respectively [37]. The high yields from macroalgae in general is attributed to the low energy required in the formation of its supporting tissue during growth. Seaweeds can also absorb nutrients across its entire surface and can be cultivated three dimensionally in water [37].

Seaweeds are composed of carbohydrates, proteins, lipids and minerals which ranges from 30 to 60%, 10–40%, 0.2–3% and 10–40%, respectively [38]. Besides their unique and varying composition, seaweeds have been grouped into three, based on their pigmentation. They are rhodophyceae (red seaweeds), phaeophyceae (brown seaweeds) and chlorophyceae (green seaweeds) based on their pigments r-phycoerythrin, chlorophyll and xanthophyll, respectively [39].

Biomass type	Species	Carbohydrate	Protein	Lipid	Ash	Ref.
Macroalgae (seaweed)	*Chaetomorpha linum*	54	—	—	22	[35]
	Caulerpa lentillifera	38.7	10.4	1.1	37.2	[40]
	C. linum	29.8	8.6	2.6	30.5	[41]
	Codium fragile	58.7	15.3	0.9	25.1	[42]
	Ulva fasciata	31.3	14.4	1.5	28.0	[43]
	Ulva lactuca	54.3	20.6	6.2	18.9	[44]
	Ulva pertusa	52.3	25.1	0.1	22.5	[38]
	Ulva rigida	53	23.4	1.2	21.7	[45]
	Chondrus pinnulatus	64.4	22.5	0.2	12.9	[46]
	Cryptonemia crenulata	47	—	—	19	[47]
	Kappaphycus alvarezzi	60.7	17.4	0.8	21.1	[48]
	K. alvarezzi	55			23	[47]
	Eucheuma cottonii	26.5	9.8	1.1	46.2	[40]
	Gelidium amansii	66.0	20.5	0.2	13.3	[42]
	Gigartina tenella	42.2	27.4	0.9	24.5	[46]
	Hypnea charoides	57.3	18.4	1.5	22.8	[26]
	Hypnea musciformis	39	—	—	22	[47]
	H. musciformis	37	—	—	30	[47]
	Hydropuntia dentata	31.2	10.3	3.2	38.7	[43]
	Lomentaria hakodatensis	40.4	29	0.7	29.9	[46]
	L. digitata	64.2	3.1	1.0	11.9	[49]
	Laminaria japonica	51.9	14.8	1.8	31.5	[44]
	Sargassum fulvellum	39.6	13	1.4	46	[44]
	Sargassum polycystum	33.5	5.4	0.3	42.4	[40]
	Sargassum vulgare	32.6	10.3	1.0	27.2	[43]
	Saccharina latissima	16.8	10.1	0.5	34.6	[49]

Biomass type	Species	Carbohydrate	Protein	Lipid	Ash	Ref.
Microalgae	*Scenedesmus acutus*	39.0	8.0	41.0	2.0	[50]
	Scenedesmus obliquus	25.0	48.8	22.5	12.9	[51]
	Pseudochoricystis ellipsoidea	19.3	27.5	45.4	2.3	[51]
	Chlorogloeopsis fritschii	37.8	41.8	8.2	4.6	[51]
	Chlorella vulgaris	16.7	41.0	10.0	13.4	[52]
	Chlorella emersonii	37.9	9.0	29.3	2.8	[51]
	Chlorella zofingiensis	11.5	11.2	56.7	4.8	[51]
	Spirulina sp.	15.1	50.1	12.3	7.6	[51]
	Nannochloropsis sp.	37.3	32.2	25.0	5.5	[53]
	Schizochytrium limacinum	25.3	12.4	56.7	5.6	[53]
	Chlorella vulgaria	43.4	28.2	17.9	10.5	[53]
	Scenedesmus sp.	35.4	24.6	10.5	29.5	[53]
	Chlamydomonas reinhardtii	35.5	34.2	24.2	6.1	[53]
	Dunaliella tertiolecta	21.7	61.3	2.9	13.5	[54]
	Botryococcus braunii	2.4	39.6	33.0	7.5	[52]
	Spirulina platensis	11.0	42.3	11.0	7.1	[52]
	Chaetoceros muelleri	34.2	16.3	43.4	—	[51]

Table 2.
Composition of typical algal biomass used in biorefinery applications.

Algal biomass composition has been found to vary based on several factors such as the season, availability of nutrients, water salinity and availability of sunlight (**Table 2**) [55]. The algal component of primary importance to bioethanol production is the carbohydrates (polysaccharides), since they currently form the only fraction that can be fermented to ethanol. Generally, some algae are composed of large fractions of complex sulphated polysaccharides which are uniquely different in each group serving as their cellular storage and structural support tissue [56]. The composition of typical algal biomass that have been considered for various biorefinery applications are presented in **Table 2**.

4. Processes for bioethanol production

The conversion of cellulosic and algal biomass to bioethanol usually involves four major processes excluding biomass selection. They include biomass pretreatment, hydrolysis of pretreated biomass, fermentation of biomass hydrolysates and ethanol recovery from the fermentation broth using distillation and dehydration processes [46]. The various efficiencies of each process will influence the final ethanol yield therefore each process condition and catalyst used is carefully selected and in most cases optimised to maximise the process efficiencies.

One of the most influential processes in bioethanol production from cellulosic and algal biomass is pretreatment. This process is used to render biomass susceptible to further breakdown by separating the cellulose, hemicellulose and lignin fractions. The selection of an efficient and cost effective biomass pretreatment method has been a major hurdle in cellulosic bioethanol production and its commercialisation for several decades. Different pretreatment mechanisms have been developed with varying degrees of efficiency [20]. All these methods have been developed with a common aim of finding a good balance between efficiency, cost, environmental effects and energy use. So far, all the methods developed have come with intrinsic advantages and disadvantages. Some common disadvantages experienced include: degradation of sugars, formation of inhibitors, high energy requirements, catalyst requirements, difficulties in catalyst recovery, challenges in waste treatment and high overall costs [20]. One or more these drawbacks are experienced in the various pretreatment processes currently developed. Nonetheless, a careful comparison and risk analysis could be used to distinguish and select one from the other. The biomass specificity for particular pretreatments could be explored to see the variations in the interactions between various cellulosic and algal biomass and various pretreatment methods as a solution.

The hydrolysis process in bioethanol production is one of the most limiting stages in the entire production process since it is the stage where the sugars to be converted to ethanol is obtained. Hydrolysis simply refers to cleavage or division through the addition of water molecules. In the context of complex sugars (polysaccharides), it involves the use of a water molecule by a catalyst to break the glycosidic linkages within their polymeric form (di-, tri-, oligo- or polysaccharide) to their monomeric form (monosaccharides or reducing sugars). During the cleavage of sugars, a hydrogen atom (H^+) is gained by one part of the polymeric structure whiles the other gains a hydroxyl group (OH^-). Thus, the separation continues until all polymeric units are reduced to their individual monomeric form [46].

The hydrolysis of cellulosic biomass for bioethanol production involves the breakdown of polymeric units such as cellulose and hemicellulose whiles the hydrolysis of algal biomass (particularly macroalgae) involves the breakdown of polymeric units such as laminarin, ulvan, alginate, carrageenan, mannitol, agar and cellulose. The simple sugars (monosaccharides) recovered from both agal and cellulosic biomass include glucose, galactose, rhamnose, mannose, fucose, xylose and arabinose for fermentation to ethanol [57]. The common methods that have been used in cellulosic and algal biomass hydrolysis includes dilute acid thermal [58], dilute alkaline thermal [59], enzymatic [3] and thermal [58] hydrolysis. All other hydrolysis methods are usually derivatives of these and are usually broadly grouped under physical, chemical, thermal and biological hydrolysis. Two or more of these methods are often combined to improve the efficiency of monomeric sugar recovery.

Enzymatic hydrolysis, particularly the use of cellulases in both cellulosic and algal biomass hydrolysis, has been promoted extensively over all other forms of hydrolysis. This is because enzymes are considered more environmentally friendly in their application and generate no inhibitors as is the case with chemical catalysts. Three major cellulase activity systems have been identified to be involved in cellulosic hydrolysis. The enzymes involved in these systems include endoglucanases, exoglucanases (cellodextrinases) and β-glucosidases [60]. Cellulase synthesis is predominant among fungi such as *Trichoderma reesei*, *Aspergillus niger*, and *Humicola insolens*; and bacteria such as *Bacillus subtilis*, *Streptomyces drodowiczi*, and *Bacillus pumilusand* [20]. Studies in enzymatic hydrolysis have focused on process optimization, improving cellulase activities, optimisation of reaction conditions, enzyme-to-substrate ratios and enzyme recovery and reuse strategies. The ideal final enzyme or enzyme cocktail should have high hydrolytic efficiencies on the preferred biomass, operate at mildly acidic or alkaline pH, be resilient to process stresses and be cost-effective [30].

The fermentation process in bioethanol production is the stage within which the reducing sugars obtained after hydrolysis are converted to ethanol by an organism. This process is always dependent on the overall ethanol production pathway selected. Currently, the ethanol pathways that have been used in cellulosic and algal biomass processing include: separate hydrolysis and fermentation (SHF), simultaneous saccharification and fermentation (SSF), simultaneous saccharification and co-fermentation (SSCF) and consolidated biomass processing (CBP) [46]. SHF is the most common and most well-developed approach which allows the use of the optimal conditions for both the hydrolysis and fermentation processes [61]. It offers the flexibility of choosing various hydrolysis processes, a feature which cannot be found in the use of the SSF approach. The SSF process involves the co-application of the enzyme for saccharification and the organism for fermentation to the pretreated biomass in the reactor under similar conditions of operation. This process is considered more cost-effective than SHF but comparisons on its process efficiency relative to SHF is currently inconclusive [46].

5. Integrated biorefinery applications to lignocellulosic biomass

Processing of cellulosic biomass using the biorefinery approach has often had its roots in the processing of first generation biomass. Typical first generation biomass such as corn, sugarcane and cassava (mostly in Africa and Asia) are still the most preferred feedstock in commercial fermentation processes. The biorefinery way of processing corn by microbial fermentation often yields ethanol, citric acid, lactic acid or lysine as the main product depending the primary product goal of the biorefinery [62]. Conventionally, the starch fraction of the corn is processed to dextrose via enzymatic pathways before microbial fermentation to the desired product. Corn fibre, gluten meal and corn steep liquor are the usual by-products obtained in a corn biorefinery which are of enormous value. Corn fibre which is lignocellulosic in nature can be further hydrolysed to obtain glucose, xylose and other monomeric sugars which can be further fermented to products such as ethanol, xylitol and acetate [63]. Gluten meal from corn which is very high in proteins can be used as feed for livestock and poultry or as substrates for various pharmaceutical products and commercial polymers [64]. Corn steep liquor which is also high in proteins is often used as a nitrogen source in various fermentation processes [65].

Sugarcane biorefineries are usually very interesting due to the unique composition of sugarcane which is usually 11–16% sucrose, 70–75% water and 10–16% fibre [63]. Sugarcane processing begins with the extraction of cane juice which immediately leads to the generation of solid residue in the form of sugarcane bagasse. Sugarcane bagasse is very high in fibre and is considered a lignocellulosic biomass. This bagasse can be valorised in a relatively more complex pathway to ethanol and other chemicals using microbial fermentation technologies or simply used as fuel in boilers for the generation of steam and electricity. The latter is the predominant process application of bagasse in industry currently. Sugarcane alone as a single biomass can be processed to obtain first generation ethanol from the cane juice and second generation ethanol from the bagasse. Additionally, sugar processing plants which use sugarcane obtain molasses as a sucrose-rich by-product which can also be used as substrate for ethanol production [63].

The potential co-production of ethanol and xylitol from sugarcane bagasse was examined in a study by Unrean and Ketsub [66]. In the study, cellulose and hemicellose fractions of the sugarcane bagasse were separated using sulphuric acid and enzymatic hydrolysis processes. The pretreated cellulose was used as substrate for the recovery of ethanol using *Saccharyomyces cerevisiae* as the fermenting organism while hemicellulose hydrolysate was used as substrate for the recovery of xylitol with *Candida tropicalis* as the fermenting organism. The product recoveries reported from the use of the bagasse was 0.44 g/g total glucose for ethanol and 0.50 g/g total xylose for xylitol. An economic analysis within the same study revealed a 2.3 fold increase in profitability for the integrated ethanol and xylitol production process over standalone cellulosic ethanol production [66].

Cellulosic pulp and paper mill waste in the form of primary sludge was examined in a study as a substrate for the production of bioethanol and biolipids [67]. In the integrated study, bioethanol and biolipids were both obtained from the hydrolysates of the primary sludge at yields of 9% and 37.8%, respectively. *S. cerevisiae* was used as the fermenting organism for bioethanol while the oleaginous yeast *Cutaneotrichosporon oleaginosum* was used as the organism for the biolipids production. A unique addition to the biorefinery process was the use of the unhydrolysed primary sludge as a cement additive or fibre reinforcement material in comparison with conventional Portland cement. The comparison of the compression load between the two materials indicated that the unhydrolysed paper mill material had 102% higher compressive strength than the Portland cement [67]. This unique application of fermentation based and non-fermentation based processes to harness the use of the pulp and paper residual biomass in a zero waste approach can be explored for other lignocellulosic biomass.

Dairy manure, a nitrogen rich cellulosic biomass has been examined as a substrate for the co-production of fumaric acid and chitin [68]. Fumaric acid is commonly used in food flavouring and preservation while chitin is a natural biopolymer with applications in the water treatment and pharmaceutical industries. In the study by Liao et al. [68], *Rhizopus oryzae* ATCC 20344 was applied as fermenting organism in a one-pot fermentation process to obtain fumaric acid in the liquid medium of the broth while chitin was found in the resulting fungal biomass formed in the broth. A maximum fumaric acid yield of 31% and a chitin yield of 0.21 g/g fungal biomass (from 11.5 g/l fungal biomass concentration) was obtained [68].

Wheat straw and corn stover have been studied as substrates for the co-production of hydrogen and ethanol [69]. In the study, genetically engineered *Escherichia coli* were applied in a dark fermentation process as means to maximise the simultaneous production of the two products. The engineered strained of *E. coli* produced a 30% increase in the co-production yield of hydrogen and ethanol. The yields obtained were 323 ml H_2/g total reducing sugars (TRS) and 3.5 g ethanol/g TRS for wheat straw and; 337.1 ml H_2/g TRS and 2.9 g ethanol/g TRS for corn stover [69].

Lignin utilisation has been a very important of part of the goal to maximise the use of lignocellulosic biomass in a biorefinery context. Considerably large quantities of lignin-rich by-products are generated from the conversion of cellulosic biomass to biofuels and various organic compounds. The efficient use of the lignin can improve the overall economics of the commercial use of lignocellulosic biomass [70]. A wide range of polymeric materials which can be used as precursors for even more valuable products have successfully been derived from lignin. They include polyesters, epoxy and phenolic resins, hydrogels, graft polymers, vanillin and polyamides. Vanillin in particular is an important compound used as flavouring agent in the food and pharmaceutical industries. It has also been considered as a precursor to hydrogels, polyester epoxide and polyethylene. Direct lignin recovery from lignocellulosic biomass can be effected using the Kraft process, lignosulfonates process, organosolv process, steam explosion or using ligninolytic enzymes such as lignin peroxidase, manganese peroxidase and laccase [70].

6. Integrated biorefinery applications to algal biomass

Several studies have used the integrated biorefinery approach to maximise the use of algal biomass and improve both their economic and process sustainability. This approach was used in the processing of the green seaweed, *C. linum* to co-produce bioethanol and biogas in a single study [41]. A bioethanol yield of 0.41 g/g reducing sugar (0.093 g/g pretreated seaweed) was obtained after the pretreatment, enzymatic hydrolysis and fermentation of the seaweed biomass. The enzymatic hydrolysis was done with a crude enzyme from *Aspergillus awamori* at 45°C and pH 5 for 30 hours while the fermentation was done with *S. cerevisiae* at 28°C for 48 hours while shaking at 150 rpm. The fermentation broth was then distilled to recover the ethanol while the residue referred to as vinasse was used as the feed for anaerobic digestion. The anaerobic digestion of the vinasse which was done at 38°C in a 0.5 l digester for 30 days yielded 0.26 l/g VS of biomethane [41]. The final waste generated was 0.3 g/g biomass which represents a substrate utilisation of up to 70%. This approach did indeed enhance the use of the substrate.

Ashokkumar *et al.* [71] also made a similar attempt with the biorefinery approach. They considered the integrated conversion of the brown seaweed *Padina tetrastromatica* to both biodiesel and bioethanol. The crude lipids content was first extracted from the biomass using various solvents to obtain a yield of 8.15% w/w biomass. This was processed further through transesterification (the process of exchanging the organic group R' of an ester with the organic group R' of an alcohol) to obtain a final biodiesel yield of 78 mg/g biomass. The residual biomass after lipids extraction was hydrolysed and fermented using baker's yeast to obtain a bioethanol yield of 161 mg/g residual biomass [71]. This study demonstrated that the integration of biodiesel and bioethanol production processes on a single seaweed biomass can efficiently harness both the lipid and carbohydrate fraction which could form up to 70% of the entire biomass.

A unique application of the biorefinery approach was used by Xu *et al.* [72]. In their study, mannitol was first removed from the brown seaweed *L. japonica* leaving behind an alginate rich suspension. The alginate suspension was used as substrate for volatile fatty acid (VFA) production via fermentation. The VFAs produced were recombined with the mannitol to produce lipids through fermentation with the oleaginous yeast, *Cryptococcus curvatus*. During the alginate fermentation process several by-products were obtained including; acetate, succinate, lactate, formate, propionate, butyrate and ethanol. A maximum lipids yield of 48.3% was achieved. The lipids obtained were very high in oleic acid (48.7%), palmitic acid (18.2%) and linoleic acid (17.5%) which indicates a fatty acids composition similar to vegetable oil [72]. The lipids can therefore be used for a myriad of applications including culinary processes and biodiesel production.

Dong *et al.* [50] were able to effectively hydrolyze the microalgae *S. acutus* to obtain reducing sugars while making the lipids more easily extractable. An ethanol concentration of 22.7 g/l was obtained from the algae while the recovery of lipids was in the range of 82–87% of total lipids after ethanol removal [50]. There was no adverse effect observed on lipids recovery due to either the acid pretreatment or the fermentation of soluble sugar processes which preceded the lipids extraction. The fatty acid methyl esters concentration was also found to be high for the lipids recovered which makes it a good substrate for biodiesel production. Lee *et al.* [73] also recovered similar products of lipids and ethanol from the microalgae, *D. tertiolecta*. In their study, 48 g lipids were extracted from 220 g of the microalgae while the residual biomass after lipid extraction was found to have a carbohydrates content of 51.9%. Upon fermentation with *S. cerevisiae*, 0.14 g ethanol/g residual biomass (0.44 g ethanol/g glucose) was obtained from the residual biomass [73]. The successful demonstration of potential biodiesel and bioethanol co-production from microalgae indicatives high potential improvements in the economic feasibility of microalgal biorefineries.

The red macroalgae, *Gracilaria verrucosa* was used a substrate for the co-production of agar (a hydrocolloid) and ethanol [74]. In the study, 33% agar was extracted from the biomass while the residual pulp was enzymatically hydrolysed to obtain 0.87 g reducing sugars/g cellulose. The hydrolysate obtained from the pulp was fermented with *S. cerevisiae* to produce ethanol with a yield of 0.43 g/g reducing sugars. A mass balance assessment in the study indicated that for every 1000 kg of dried algal biomass, 280 kg of agar can be obtained together with 38 kg of ethanol. Additionally, 20 and 25 kg of lipid and protein, respectively can be obtained from the residual pulp after agar extraction [74]. In a similar approach, the hydrocolloid, carrageenan was first extracted from the seaweed *E. cottonii* before the application of the residual pulp in ethanol production [3]. The carrageenan extraction led to an increase in the cellulose fraction to 64% in the residual seaweed pulp. Ethanol yields of 0.25–0.27 g/g residual seaweed pulp were obtained using *S. cerevisiae* as the fermenting organism [3].

Co-production of biosolar hydrogen and biogas was explored on the microalgae *C. reinhardtii* as a means to evaluate the integrated biorefinery approach to processing the biomass [75]. Hydrogen was first produced using the sulphur deprivation method. This method involves the cultivation of algal cells in a sulphur-containing medium until the cells reach the stationary growth phase. Cell pellets are then harvested and re-suspended in sulphur-free medium followed by incubation in light at 600 $\mu mol/m^2/s$ under room temperature. The production of hydrogen prior to anaerobic digestion of the microalage resulted in a 123% increase in biogas generation from an initial 587 ml biogas/g volatile solids with 66% CH_4 content [75].

In another biorefinery process, the microalgae *Nannochloropsis* sp. was used as substrate for the recovery of three different valuable products [76]. Supercritical CO_2 was used to extract 45 g lipids/100 g dry biomass and 70% of pigments which were mainly carotenoids. The residual microalgal biomass after extraction was used as an efficient substrate to produce hydrogen at a yield of 60.6 ml/g dry biomass through dark fermentation with *Enterobacter aerogenes* [76]. Harnessing these valuable products from a single biomass shows high economic prospects for microalgal biorefineries.

7. Additional prospects for biorefinery applications

Prospects in other biomass conversion pathways such as thermochemical, mechanical and chemical cannot be completely ignored and in some cases entirely replaced with the biochemical processes proposed (**Figure 1**). Thermochemical processes such gasification which involves the application of heat to biomass at high temperatures (> 700°C) in the presence of low oxygen concentrations can be used to obtain syngas (mixture of methane, hydrogen, carbon dioxide and carbon monoxide) [77]. Syngas can be used as a standalone fuel or a platform chemical for the production of alcohols and organic acids. Alternatively, biomass can be subjected to a pyrolysis process which involves the use of temperatures between 300 and 600°C in the absence of oxygen to convert the biomass to a liquid bio-oil with biochar and light gases as by-products [78]. Such thermochemical processes could be considered as downstream processes after lignocellulosic and algal biomass fermentation where large non-cellulose fractions are generated as side-streams. A variant thermochemical process is hydrothermal treatment or upgradation. It involves the use of high temperature (200–600°C) and pressure (5–40 MPa) liquids often in the form of supercritical water to produce various liquid fuels [33].

Mechanical processes which do not typically change the composition of biomass but tend to reduce sizes or separate impurities or other components are usually applied in most biorefinery processes. It is particularly popular when handling and pre-treating lignocellulosic biomass [79]. However, there are mechanical processes that are considered complete standalone processes which generates their own useful products. A typical example is briquetting. Briquettes are often in the form of relatively evenly sized pellets produced by the compression of carbon-rich biomass. They are known to burn longer and produce a lower net greenhouse gas emissions which promotes their use as good substitutes to coal, charcoal and raw firewood [80]. Such a process could be used as a downstream process after lignocellulosic and algal biomass fermentation to minimise waste generation and add more value to residual materials.

8. Sustainability and circular economy perspectives of cellulosic and algal biorefineries

Circular economies principally emphasise the development of economic systems that eliminate waste and continuously utilise resources. In the context of biomass resources, an alternative term often used is Circular bioeconomy. Biomass is emerging as the primary renewable resource to tackle several challenges especially with regards to greenhouse gas emissions and depleting fossil fuels [6]. Therefore several technologies and multi-technology integration systems are being promulgated as the backbone for a Circular bioeconomy. The European Union describes this Circular bioeconomy as one that encompasses the formation of various renewable biological resources and their conversion to several high-value bio-based products such as food, feed, chemicals, and energy [81]. At the heart of this economic model is the biorefinery concept which has been elaborately described in this review. The biorefinery concept's role especially for algal and lignocellulosic biomass processing is to optimise the conversion of these biomass to achieve the goals principally set for the circular bioeconomy [82]. Lignocellulosic biomass utilisation will be key to the success of the bioeconomy because they are the primary components of most biological wastes generated especially from crop production and processing. The unique benefits derived from the use of algal biomass in particular includes no arable land requirements, high biomass productivity and no reliance on fresh water and fertiliser sources [2]. This makes it an equally important resource for the circular bioeconomy.

The circular bioeconomy and the circular economy in a broader context have direct positive ripple effects on the social, economic and environmental concerns associated with current economic development models. These three aspects of any development process form the pillars of sustainability. It is therefore nearly impossible to dissociate the circular economy from sustainability. The role of lignocellulosic and algal biorefineries in sustainable development can be found directly in a number of the Sustainability Development Goals (SDGs) proposed by the United Nations. They include: *Zero hunger* (Goal 2) through the provision of affordable feed for livestock farming; *Clean water and sanitation* (Goal 6) through the utilisation of algal blooms which forms a major health hazard for coastal communities; *Affordable and clean energy* (Goal 7) through the conversion of cellulosic and algal biomass to biofuels; *Decent work and economic growth* (Goal 8) through the creation of small and medium scale biorefinery businesses and employment opportunities; *Industry, innovation and infrastructure* (Goal 9) through the creation of new and innovative co-product pathways using the biorefinery approach; *Sustainable cities and communities* (Goal 11) through energy recovery from the biodegradable fractions of municipal solid wastes; *Responsible Consumption and Production* (Goal 12) through the multi-product recovery from the same biomass leading to a reduction in waste fractions and; *Climate Action* (Goal 13) through the reduction in greenhouse gas emissions from crop production residue decay and direct combustion [83].

A reduction or absence of waste streams especially for agro residual biomass which is promoted by Goal 12 of the SDGs is a direct attribute of the zero waste concept. This concept refers to the design and management of products and processes in a systematic form to avoid and eliminate waste, and to recover all resources from the waste stream [84]. Resource recovery from waste streams is the primary point of intersection between the integrated biorefinery concept and the zero waste concept. The utilisation of cellulosic agro residues such stalks from various cereals reduces the apparent greenhouse gas emissions from their decay or direct combustion. This forms a simple yet effective climate change mitigation measure for both developed and developing countries.

9. Conclusions

The studies described in this chapter have highlighted the considerable benefits from the use of integrated processing technologies on lignocellulosic and algal biomass. The most obvious feature is the increased use of the substrate and the minimization of waste generated. The less obvious feature is the improvements in the economic sustainability of commercial cellulosic and algal biorefineries. These studies show that the potential range of products including fuels, chemicals and polymers that current and future biorefineries could produce is currently very extensive. Research and development efforts are adding almost daily to products and co-products of known fermentation-based biorefinery pathways. The most important consideration which has pushed research even further is the importance attached to the sustainability of processes in recent years. Sustainability is now an equally important consideration in addition to economic feasibility, product yield, process efficiency and selectivity. This is due to the importance of developing climate smart yet cost-effective technologies and processes which will protect and preserve ecosystems for present and future generations. The integrated biorefinery approach has therefore become indispensable to productive and sustainable biomass processing.

Acknowledgements

The author is grateful to Moses Mensah, Robert Aryeetey and George vanDyck for their continuous technical support.

Conflict of interest

The author declares no conflict of interest.

Author details

Felix Offei
Department of Marine Engineering, Regional Maritime University, Accra, Ghana

*Address all correspondence to: felix.offei@rmu.edu.gh

References

[1] Sawin J, Rutovitz J, Sverrisson F. Advancing the global renewable energy transition. Tech. rep; 2018.

[2] Kraan S. Mass-cultivation of carbohydrate rich macroalgae, a possible solution for sustainable biofuel production. Mitigation and Adaptation Strategies for Global Change. 2013 Jan;18(1):27-46.

[3] Tan IS, Lee KT. Enzymatic hydrolysis and fermentation of seaweed solid wastes for bioethanol production: An optimization study. Energy. 2014 Dec 15;78:53-62.

[4] Gavrilescu M. Biorefinery systems: an overview. Bioenergy research: advances and applications. 2014 Jan 1:219-241.

[5] Pande M, Bhaskarwar AN. Biomass conversion to energy. In Biomass conversion 2012 (pp. 1-90). Springer, Berlin, Heidelberg.

[6] Cherubini F. The biorefinery concept: using biomass instead of oil for producing energy and chemicals. Energy conversion and management. 2010 Jul 1;51(7):1412-1421.

[7] Maeda RN, Serpa VI, Rocha VA, Mesquita RA, Santa Anna LM, De Castro AM, Driemeier CE, Pereira Jr N, Polikarpov I. Enzymatic hydrolysis of pretreated sugar cane bagasse using Penicillium funiculosum and Trichoderma harzianum cellulases. Process Biochemistry. 2011 May 1;46(5):1196-1201.

[8] García-Aparicio MP, Oliva JM, Manzanares P, Ballesteros M, Ballesteros I, González A, Negro MJ. Second-generation ethanol production from steam exploded barley straw by Kluyveromyces marxianus CECT 10875. Fuel. 2011 Apr 1;90(4):1624-1630

[9] Abbas A, Ansumali S. Global potential of rice husk as a renewable feedstock for ethanol biofuel production. BioEnergy Research. 2010 Dec;3(4):328-334.

[10] Wang L, Yang M, Fan X, Zhu X, Xu T, Yuan Q. An environmentally friendly and efficient method for xylitol bioconversion with high-temperature-steaming corncob hydrolysate by adapted Candida tropicalis. Process Biochemistry. 2011 Aug 1;46(8):1619-1626

[11] Liu CZ, Cheng XY. Improved hydrogen production via thermophilic fermentation of corn stover by microwave-assisted acid pretreatment. International Journal of Hydrogen Energy. 2010 Sep 1;35(17):8945-8952.

[12] Gullón B, Yáñez R, Alonso JL, Parajó JC. Production of oligosaccharides and sugars from rye straw: a kinetic approach. Bioresource technology. 2010 Sep 1;101(17):6676-6684.

[13] Lu X, Zhang Y, Angelidaki I. Optimization of H2SO4-catalyzed hydrothermal pretreatment of rapeseed straw for bioconversion to ethanol: focusing on pretreatment at high solids content. Bioresource technology. 2009 Jun 1;100(12):3048-3053.

[14] Talebnia F, Karakashev D, Angelidaki I. Production of bioethanol from wheat straw: an overview on pretreatment, hydrolysis and fermentation. Bioresource technology. 2010 Jul 1;101(13):4744-4753.

[15] Yadav KS, Naseeruddin S, Prashanthi GS, Sateesh L, Rao LV. Bioethanol fermentation of concentrated rice straw hydrolysate using co-culture of Saccharomyces cerevisiae and Pichia stipitis. Bioresource technology. 2011 Jun 1;102(11):6473-6478.

[16] Caparrós S, Ariza J, López F, Nacimiento JA, Garrote G, Jiménez L. Hydrothermal treatment and ethanol pulping of sunflower stalks. Bioresource technology. 2008 Mar 1;99(5): 1368-1372.

[17] Goshadrou A, Karimi K, Taherzadeh MJ. Bioethanol production from sweet sorghum bagasse by Mucor hiemalis. Industrial Crops and Products. 2011 Jul 1;34(1):1219-1225.

[18] Panichelli L, Gnansounou E. Estimating greenhouse gas emissions from indirect land-use change in biofuels production: concepts and exploratory analysis for soybean-based biodiesel. 2008.

[19] Monsalve JF, Medina de Perez VI, Ruiz Colorado AA. Ethanol production of banana shell and cassava starch. Dyna. 2006 Nov;73(150):21-27.

[20] Menon V, Rao M. Trends in bioconversion of lignocellulose: biofuels, platform chemicals & biorefinery concept. Progress in energy and combustion science. 2012 Aug 1;38(4):522-550.

[21] Jeon YJ, Xun Z, Rogers PL. Comparative evaluations of cellulosic raw materials for second generation bioethanol production. Letters in applied microbiology. 2010 Nov;51(5):518-524.

[22] nee'Nigam PS, Gupta N, Anthwal A. Pre-treatment of agro-industrial residues. InBiotechnology for agro-industrial residues utilisation 2009 (pp. 13-33). Springer, Dordrecht.

[23] Lee DH, Cho EY, Kim CJ, Kim SB. Pretreatment of waste newspaper using ethylene glycol for bioethanol production. Biotechnology and Bioprocess Engineering. 2010 Dec 1;15(6):1094-1101.

[24] Peng L, Chen Y. Conversion of paper sludge to ethanol by separate hydrolysis and fermentation (SHF) using Saccharomyces cerevisiae. Biomass and Bioenergy. 2011 Apr 1;35(4):1600-1606.

[25] Pires EJ, Ruiz HA, Teixeira JA, Vicente AA. A new approach on brewer's spent grains treatment and potential use as lignocellulosic yeast cells carriers. Journal of agricultural and food chemistry. 2012 Jun 13;60(23):5994-5999.

[26] Malherbe S, Cloete TE. Lignocellulose biodegradation: fundamentals and applications. Reviews in Environmental Science and Biotechnology. 2002 Jun;1(2): 105-114.

[27] Cara C, Ruiz E, Oliva JM, Sáez F, Castro E. Conversion of olive tree biomass into fermentable sugars by dilute acid pretreatment and enzymatic saccharification. Bioresource technology. 2008 Apr 1;99(6): 1869-1876.

[28] Aswathy US, Sukumaran RK, Devi GL, Rajasree KP, Singhania RR, Pandey A. Bio-ethanol from water hyacinth biomass: an evaluation of enzymatic saccharification strategy. Bioresource technology. 2010 Feb 1;101(3):925-930.

[29] Shafiei M, Karimi K, Taherzadeh MJ. Pretreatment of spruce and oak by N-methylmorpholine-N-oxide (NMMO) for efficient conversion of their cellulose to ethanol. Bioresource Technology. 2010 Jul 1;101(13): 4914-4918.

[30] Knauf M, Moniruzzaman M. Lignocellulosic biomass processing: a perspective. International sugar journal. 2004;106(1263):147-150.

[31] Arantes V, Saddler JN. Access to cellulose limits the efficiency of enzymatic hydrolysis: the role of amorphogenesis. Biotechnology for biofuels. 2010 Dec;3(1):1-1.

[32] Adams JM, Ross AB, Anastasakis K, Hodgson EM, Gallagher JA, Jones JM, Donnison IS. Seasonal variation in the chemical composition of the bioenergy feedstock Laminaria digitata for thermochemical conversion. Bioresource technology. 2011 Jan 1;102(1):226-234.

[33] Singh R, Bhaskar T, Bala gurumurthy B. Hydrothermal upgradation of algae into value-added hydrocarbons. In Biofuels from Algae 2014 Jan 1 (pp. 235-260). Elsevier.

[34] Ab Kadir MI, Ahmad WW, Ahmad MR, Misnon MI, Ruznan WS, Jabbar HA, Ngalib K, Ismail A. Utilization of eco-colourant from green seaweed on textile dyeing. In Proceedings of the International Colloquium in Textile Engineering, Fashion, Apparel and Design 2014 (ICTEFAD 2014) 2014 (pp. 79-83). Springer, Singapore.

[35] Schultz-Jensen N, Thygesen A, Leipold F, Thomsen ST, Roslander C, Lilholt H, Bjerre AB. Pretreatment of the macroalgae Chaetomorpha linum for the production of bioethanol–Comparison of five pretreatment technologies. Bioresource Technology. 2013 Jul 1;140:36-42.

[36] FAO, Year book of Fishery Statistics. Available from: ftp://ftp.fao.org/FI/STAT/summary/default.htm. 2013. Accessed 28 January 2016

[37] Adams JM, Gallagher JA, Donnison IS. Fermentation study on Saccharina latissima for bioethanol production considering variable pre-treatments. Journal of applied Phycology. 2009 Oct;21(5):569-574.

[38] Lee SY, Chang JH, Lee SB. Chemical composition, saccharification yield, and the potential of the green seaweed Ulva pertusa. Biotechnology and bioprocess engineering. 2014 Nov 1;19(6): 1022-1033.

[39] Borines MG, De Leon RL, McHenry MP. Bioethanol production from farming non-food macroalgae in Pacific island nations: Chemical constituents, bioethanol yields, and prospective species in the Philippines. Renewable and Sustainable Energy Reviews. 2011 Dec 1;15(9):4432-4435.

[40] Matanjun P, Mohamed S, Mustapha NM, Muhammad K. Nutrient content of tropical edible seaweeds, Eucheuma cottonii, Caulerpa lentillifera and Sargassum polycystum. Journal of Applied Phycology. 2009 Feb;21(1):75-80.

[41] Yahmed NB, Jmel MA, Alaya MB, Bouallagui H, Marzouki MN, Smaali I. A biorefinery concept using the green macroalgae Chaetomorpha linum for the coproduction of bioethanol and biogas. Energy Conversion and Management. 2016 Jul 1;119:257-265.

[42] Hong IK, Jeon H, Lee SB. Comparison of red, brown and green seaweeds on enzymatic saccharification process. Journal of Industrial and Engineering Chemistry. 2014 Sep 25;20(5):2687-2691.

[43] Offei F, Mensah M, Kemausuor F. Cellulase and acid-catalysed hydrolysis of Ulva fasciata, Hydropuntia dentata and Sargassum vulgare for bioethanol production. SN Applied Sciences. 2019 Nov;1(11):1-3.

[44] Kim NJ, Li H, Jung K, Chang HN, Lee PC. Ethanol production from marine algal hydrolysates using Escherichia coli KO11. Bioresource technology. 2011 Aug 1;102(16): 7466-7469.

[45] El Harchi M, Kachkach FF, El Mtili N. Optimization of thermal acid hydrolysis for bioethanol production from Ulva rigida with yeast Pachysolen tannophilus. South African Journal of Botany. 2018 Mar 1;115:161-169.

[46] Offei F, Mensah M, Thygesen A, Kemausuor F. Seaweed bioethanol production: A process selection review on hydrolysis and fermentation. Fermentation. 2018 Dec;4(4):99.

[47] Rhein-Knudsen N, Ale MT, Ajalloueian F, Yu L, Meyer AS. Rheological properties of agar and carrageenan from Ghanaian red seaweeds. Food Hydrocolloids. 2017 Feb 1;63:50-58.

[48] Fayaz M, Namitha KK, Murthy KC, Swamy MM, Sarada R, Khanam S, Subbarao PV, Ravishankar GA. Chemical composition, iron bioavailability, and antioxidant activity of Kappaphycus alvarezzi (Doty). Journal of agricultural and food chemistry. 2005 Feb 9;53(3):792-797.

[49] Manns D, Deutschle AL, Saake B, Meyer AS. Methodology for quantitative determination of the carbohydrate composition of brown seaweeds (Laminariaceae). Rsc Advances. 2014;4(49):25736-25746.

[50] Dong T, Knoshaug EP, Davis R, Laurens LM, Van Wychen S, Pienkos PT, Nagle N. Combined algal processing: A novel integrated biorefinery process to produce algal biofuels and bioproducts. Algal Research. 2016 Nov 1;19:316-323.

[51] Biller P, Ross AB. Pyrolysis GC–MS as a novel analysis technique to determine the biochemical composition of microalgae. Algal Research. 2014 Oct 1;6:91-97.

[52] Sydney EB, Sturm W, de Carvalho JC, Thomaz-Soccol V, Larroche C, Pandey A, Soccol CR. Potential carbon dioxide fixation by industrially important microalgae. Bioresource technology. 2010 Aug 1;101(15):5892-5896.

[53] Yao L, Gerde JA, Lee SL, Wang T, Harrata KA. Microalgae lipid characterization. Journal of agricultural and food chemistry. 2015 Feb 18;63(6):1773-1787.

[54] Shuping Z, Yulong W, Mingde Y, Kaleem I, Chun L, Tong J. Production and characterization of bio-oil from hydrothermal liquefaction of microalgae Dunaliella tertiolecta cake. Energy. 2010 Dec 1;35(12):5406-5411.

[55] Dahiya A. Algae biomass cultivation for advanced biofuel production. InBioenergy 2015 Jan 1 (pp. 219-238). Academic Press.

[56] Chirapart A, Praiboon J, Puangsombat P, Pattanapon C, Nunraksa N. Chemical composition and ethanol production potential of Thai seaweed species. Journal of applied phycology. 2014 Apr;26(2):979-986.

[57] Lee KT, Ofori-Boateng C. Biofuels: production technologies, global profile, and market potentials. In Sustainability of biofuel production from oil palm biomass 2013 (pp. 31-74). Springer, Singapore.

[58] Yazdani P, Zamani A, Karimi K, Taherzadeh MJ. Characterization of Nizimuddinia zanardini macroalgae biomass composition and its potential for biofuel production. Bioresource technology. 2015 Jan 1;176:196-202.

[59] Trivedi N, Gupta V, Reddy CR, Jha B. Enzymatic hydrolysis and production of bioethanol from common macrophytic green alga Ulva fasciata Delile. Bioresource technology. 2013 Dec 1;150:106-112.

[60] Lynd LR, Weimer PJ, Van Zyl WH, Pretorius IS. Microbial cellulose utilization: fundamentals and biotechnology. Microbiology and molecular biology reviews. 2002 Sep 1;66(3):506-577.

[61] Kim HM, Wi SG, Jung S, Song Y, Bae HJ. Efficient approach for bioethanol production from red

seaweed Gelidium amansii. Bioresource technology. 2015 Jan 1;175:128-134.

[62] Bevan MW, Franssen MC. Investing in green and white biotech. Nature biotechnology. 2006 Jul;24(7):765-767.

[63] Hossain GS, Liu L, Du GC. Industrial bioprocesses and the biorefinery concept. In Current Developments in Biotechnology and Bioengineering 2017 Jan 1 (pp. 3-27). Elsevier.

[64] Shukla R, Cheryan M. Zein: the industrial protein from corn. Industrial crops and products. 2001 May 1;13(3):171-192.

[65] Agarwal L, Dutt K, Meghwanshi GK, Saxena RK. Anaerobic fermentative production of lactic acid using cheese whey and corn steep liquor. Biotechnology letters. 2008 Apr;30(4):631-635.

[66] Unrean P, Ketsub N. Integrated lignocellulosic bioprocess for co-production of ethanol and xylitol from sugarcane bagasse. Industrial crops and products. 2018 Nov 1;123:238-246.

[67] Zambare VP, Christopher LP. Integrated biorefinery approach to utilization of pulp and paper mill sludge for value-added products. Journal of Cleaner Production. 2020 Nov 20;274:122791.

[68] Liao W, Liu Y, Frear C, Chen S. Co-production of fumaric acid and chitin from a nitrogen-rich lignocellulosic material–dairy manure–using a pelletized filamentous fungus Rhizopus oryzae ATCC 20344. Bioresource technology. 2008 Sep 1;99(13):5859-5866.

[69] Lopez-Hidalgo AM, Sánchez A, De León-Rodríguez A. Simultaneous production of bioethanol and biohydrogen by Escherichia coli WDHL using wheat straw hydrolysate as substrate. Fuel. 2017 Jan 15;188:19-27.

[70] Ahmad E, Pant KK. Lignin conversion: a key to the concept of lignocellulosic biomass-based integrated biorefinery. In Waste biorefinery 2018 Jan 1 (pp. 409-444). Elsevier.

[71] Ashokkumar V, Salim MR, Salam Z, Sivakumar P, Chong CT, Elumalai S, Suresh V, Ani FN. Production of liquid biofuels (biodiesel and bioethanol) from brown marine macroalgae Padina tetrastromatica. Energy conversion and Management. 2017 Mar 1;135:351-361.

[72] Xu X, Kim JY, Oh YR, Park JM. Production of biodiesel from carbon sources of macroalgae, Laminaria japonica. Bioresource technology. 2014 Oct 1;169:455-461.

[73] Lee OK, Kim AL, Seong DH, Lee CG, Jung YT, Lee JW, Lee EY. Chemo-enzymatic saccharification and bioethanol fermentation of lipid-extracted residual biomass of the microalga, Dunaliella tertiolecta. Bioresource technology. 2013 Mar 1;132:197-201.

[74] Kumar S, Gupta R, Kumar G, Sahoo D, Kuhad RC. Bioethanol production from Gracilaria verrucosa, a red alga, in a biorefinery approach. Bioresource technology. 2013 May 1;135:150-156.

[75] Mussgnug JH, Klassen V, Schlüter A, Kruse O. Microalgae as substrates for fermentative biogas production in a combined biorefinery concept. Journal of biotechnology. 2010 Oct 1;150(1):51-56.

[76] Nobre BP, Villalobos F, Barragan BE, Oliveira AC, Batista AP, Marques PA, Mendes RL, Sovová H, Palavra AF, Gouveia L. A biorefinery from Nannochloropsis sp. microalga–extraction of oils and pigments. Production of biohydrogen from the

leftover biomass. Bioresource technology. 2013 May 1;135:128-136.

[77] Spath PL, Dayton DC. Technical and economic assessment of synthesis gas to fuels and chemicals with emphasis on the potential for biomass-derived syngas. Preliminary Screening. 2003. http://handle.dtic.mil/100.2/ADA436529.

[78] Yan G, Yao W, Fei W, Yong J. Research progress in biomass flash pyrolysis technology for liquids production. Chemical Industry and Engineering Progress. 2001 Jan 1(8):13-17.

[79] Senneca O. Kinetics of pyrolysis, combustion and gasification of three biomass fuels. Fuel processing technology. 2007 Jan 1;88(1):87-97.

[80] Azasi VD, Offei F, Kemausuor F, Akpalu L. Bioenergy from crop residues: A regional analysis for heat and electricity applications in Ghana. Biomass and Bioenergy. 2020 Sep 1;140:105640.

[81] European Commission. A Sustainable Bioeconomy for Europe: strengthening the connection between economy, society and the environment, Brussels, Belgium. 2018.

[82] Ubando AT, Felix CB, Chen WH. Biorefineries in circular bioeconomy: A comprehensive review. Bioresource technology. 2020 Mar 1;299:122585.

[83] United Nations Development Programme. Transforming our world: The 2030 agenda for sustainable development. UN Summit, New York, 2015.

[84] Zaman AU, Lehmann S. The zero waste index: a performance measurement tool for waste management systems in a 'zero waste city'. Journal of Cleaner Production. 2013 Jul 1;50:123-132.

Section 2

Food Fermentations

Chapter 5

Fermentation of Bovine, Non-Bovine and Vegetable Milk

Tridjoko Wisnu Murti

Abstract

Fermented milk has been developing alongside the history of human civilization. It is observed having positive effect on gastrointestinal health. It has reaches at the steps of explaining what happens in the process, despite some information is still unclear. Fermentation involves many organisms, technique, biochemical reactions, tool and apparatus as well as cultural diversity among people and regions due to differences and changes in climate. Fermented milks, using milk as the raw material from bovine and non-bovine dairy species, and in some regions, especially in Asia and Africa, they also use materials from vegetable extracts. Some progress in Fermented Milk Science, has explained the role of such fermented foods for human health. These benefits have been more and more progressing to select specific microbes, known as probiotic cultures, which combined with specific substances from vegetable extract (prebiotic) could improve lactose digestion, role anti-cancer, anti-hypercholesteremic and anti-pathogenic bacteria as well as anti-virus were discussed in this article.

Keywords: fermented milks, probiotic and prebiotics, bovine and non-bovine milk, health benefits

1. Introduction

Fermented milk having studying since long time and having the positive effect of selected microbiota on gastrointestinal health on human from different region in the world. Fermented milk acomposted largely lactic acid bacteria with history of save use. Lactic acid bacteria are considered as the best for prebiotic-containing fermented milk while preserving the taste and nutritional properties.

This review is presented since the definition of fermented milk manufacturing source of raw milk material (Bovine, small ruminans milk as well as vegetable milk) and it's benefit role for human being.

2. Definition

Fermentation is a process involving the (re) generation of ATP by microorganisms *via* the transformation of organic substances (in this case, lactose as a sugar in the milk); specific enzymes, produced by bacteria (thermophilic, mesophilic, or intestinal bacteria) or fungi as well as yeasts, acting as catalysts in this reaction. Several categories of fermented milk, which depend on the dominant final products are: alcoholic, lactic, acetic, butyric, propionic. Advances in fermented milk production technology has led to variety of products, naming more than 400 of local fermented milk, i.e.: Kefir, yakult, yoghurt, Ymer, Lang-fii, bifidus milk, acidophilus milk, koumiss, lassi, leben, shrikhand as well as cheese, that suit diverse cultural taste. According to the *Codex alimentarius*, fermented milk is a dairy product obtained by the fermentation of milk, which may have been made from products obtained from milk with or without any modification of their composition (within certain specified limits), *via* the action of corresponding microorganisms and which result in a lowered pH with or without coagulation (isoelectric precipitation).

3. Raw milk materials

Raw materials for fermented milk are milked from different ruminants or other not only limited from cows, buffalo, sheep and goats, but also from mares as well as camel milk [1–3]. The composition of raw milk materials is in **Table 1**. Milk from Buffalo, Sheep and Yak rich in Fat contents, leads to high total solids while, Mare's milk rich in lactose as Breast milk, considered as the 1st choice to replace or to complete breastfeeding. Soymilk composition contains CP 4.5, Carbohydrate 10.0, Fat 4.3, ash 0.66 and Moisture 80.34%, respectively [6].

Type of fermentation in lactic acid bacteria used are homo- or hetero-fermentation using the main products of lactic acid for homo-fermentation bacteria or lactic acid, acetic acid, and CO_2 for hetero fermentation. The isomers of lactate are L-lactate, D-Lactate or DL-Lactate as in **Table 2** [2]. Isomer L-lactate

Source	Total solids	Fat	Total Protein	Casein	Whey Protein	Lactose	Ash
Cow	12.3–14.5	3.4–5.5	3.0–4.0	2.8	0.6	4.6–7.0	0.7
Buffalo	16.0–17.0	6.0–7.5	4.3–4.7	3.6	0.9	4.3–4.7	80.9
Goat	11.5–13.5	3.4–4.5	2.8–3.7	2.5	0.4	3.9–4.8	0.8
Sheep	16.0–20.0	6.0–8.5	5.5–6.5	4.6	0.9	4.0–4.7	1.0
Camel	13.5–16.0	5.0–5.5	3.5–4.5	2.7	0.9	5.0–6.0	0.7
Mare	10.0–12.0	1.0–2.0	1.6–1.8	1.3	1.2	6.0–7.0	0.5
Yak	17.8–18.0	6.5–9.0	5.5	NA	NA	5.0–6.0	0.9

[4, 5].

Table 1.
Milk composition of different dairy animals.

Genus	Fermentation type	Main products	Lactate isomers
Streptococcus	Homo	Lactate	L
Pediococcus	Homo	Lactate	L, DL
Lactobacillus:	Homo	Lactate	L, D or DL
• obligate homo	Homo	Lactate	L, D or DL
• Facultative hetero	Hetero	Lactate, Acetate	L, D or DL
Obligate hetero	Hetero	Lactate, Acetate, CO2	L, D or DL
Leuconostoc	Hetero	Lactate, Acetate, CO2	D
Bifidobacterium	Hetero	Lactate, acetate	L

[2].

Table 2.
Types of fermentation in lactic acid bacteria.

is more acceptable for infants. Genus bacteria, which are in fermented milks, i.e.: Lactobacillus (*L. delbrueckii subs. Bulgaricus, L. acidophilus, L. helveticus, L. brevis, L. fermenteum*, and *L. kefir*), Streptococcus (*S. thermophilus*), Leuconostoc (*Ln. mesenteroides,* and *Ln lactis*), *Pediococcus* (*P. acidilactici*, and *P. pentosaceus*), Acetobacter (*A.aceti*), and Bifidobacterium (*B. breve, B. adult, B. infantis, B. longum, B. bifidum,* and *B. pseudolongum*).

While for yeast included *Torulaspora delbrueckii, Kluyveromyces marxianus subsp. marxianus*, *Candida kefir*, and *Saccharomyces cerevisiae*. Some bacteria are considered as thermophilic or mesophilic starters, and other as intestinal bacteria, depended on optimal temperature of growth or source location [7].

4. Manufacturing of fermented milks

Characteristics of naturally fermented milk depend upon the availability of the milk in respective regions. However, fermented milk like Zabady, Laban, Rayeb, and Shubat from Northern Africa (Egypt, Morocco, and mid-west Asian countries) and Shoyu from Himalayan region as well as flmjölk and långfl from Sweden have same characteristics of fermentation [5, 7].

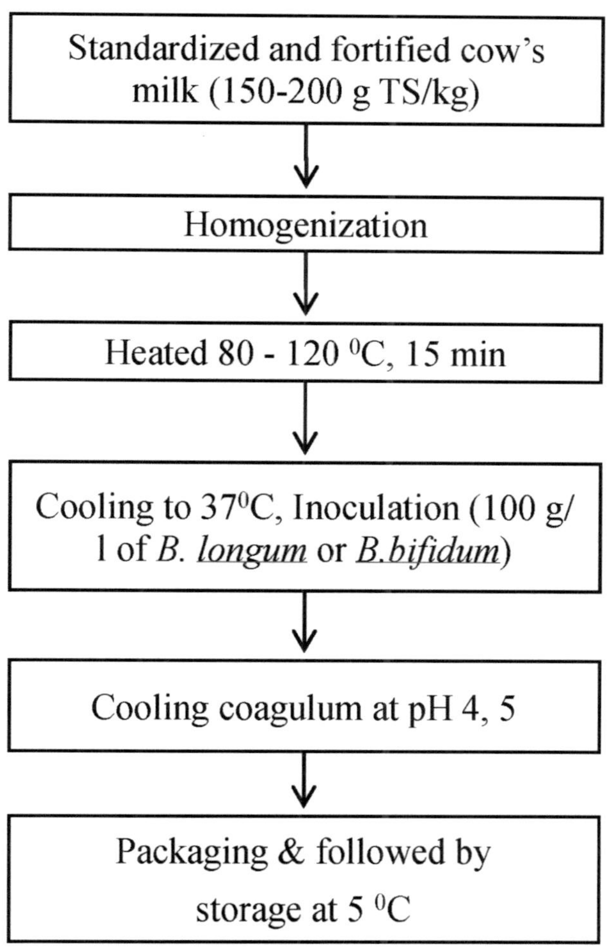

Figure 1.
General steps for manufacturing of bifidus milks.

In general, manufacturing process of fermented milk is in **Figure 1**. The technology and biochemistry of fermentation have been extensively investigated and varied among countries, regions or for special purposes. However, the principal stages of production of any type of fermented milk are included: standardization of fat content, fortification of the milk solids, homogenization, deaerization (optional), and high heat treatment have much in common reviewed. Fermented dairy products, included probiotics cultures, are generally not commercialized as in single cultures, except Yakult and Bifidus milk or acidophilus milk. However, Fermented milks contain mixed cultures.

5. Fermented non-bovine milk and vegetable extract

Fermented non-bovine milks grew along with fermented cow milk, despite its limited numbers.

Four major ruminant species (cow, goat, sheep, and buffalo) contribute raw milk for their manufacturing, on which the research has been focusing since a long time. Currently, the worldwide production of non-bovine milk reaches 133 million tons per year, accounting for more than 17% of all milk production. Of this amount, 13.5% accounts from goat milk production. Goat milk is one of the major contributors to non-bovine milk production [8]. In recent years, the social transformation underway in poor and developing countries, climatic changes, and increased attention to animal welfare are highlighting on some minor animal species that have a "local dairy concern" such as equines, camels, and yaks [9]. It has studied fermented bovine and non-bovine milk (goat and mare's milk) using raw milk or whey liquid added by probiotic bacteria, either single or multiple probiotic culture [3]. Goat milk possesses beneficial characteristics that make it potentially useful in human medicine or nutrition. Fermented Goat Milk developed in the presence of *L. casei* [10]. As well as present of *Streptococcus thermophilus* or mix probiotic cultures could well accepted by consumers [11].

The so-called vegetable milks are in the spotlight as the beverage that is lactose-free, animal protein-free and cholesterol-free features which fit well with the current demand for healthy food products [12–14]. There is an increasing demand for non-dairy probiotic foods (both fermented and non-fermented) including fruit and vegetable juices, soy and certain cereal products due to vegetarianism [15]. Soy Milk is rich in essential and branched amino acids except Met, and good source of B- vitamins especially, niacin, pyridoxine and folic acid/folacin [13]. Asia represents high demand of vegetable beverages, included fermented vegetable milk as Soymilk [12]. The food market reflects to an ever-greater degree the consumer demand for healthy food products, especially fermented beverage using cereals, nuts as almonds, peanuts, as well as fermented soymilk. France may be a country that considers, it is not milk, but extract [16]. There are two large different groups of vegetable milks or plant milks: nut and cereal milk that can be broadly classified by [12].

6. Probiotics and health benefits for human being

Fermented milks or vegetable extracts offers tremendous potential for promoting health, improving nutrition, and reducing the risks of various diseases. Lactic acid bacteria, both mesophilic and thermophilic or intestinal bacteria may be considered as probiotic cultures which improve health status of the host. Infants, children, adults, and elderly can consume fermented milks for their good taste, health benefit and their general nutritional value, reducing the incidence of malnutrition, lactose intolerance and diarrhea. Those with special medical needs can switch to fermented milks to provide added nutrition, solve intestinal disorders, improve immune function, and optimize gut ecology. Offering general health benefit effects for human indicated the role of probiotic bacteria. Modern production makes these products available to world population, resulting in the consumption of fermented milk products will reach new milestone.

Probiotics beverages belong to fermented milks are known as health promoting foods which have market estimated at USD 49.4 billion in 2018 and projected to grow reaching 69.3 billion in 2023. The increasing use of yeast *Saccharomyces boulardii*, follows a success story of probiotic containing yoghurt in China, Brazil, and India. Japan, which has increasingly aging population, consider health benefits associated with probiotic-fortified foods and beverages. Therefore, Japan may have new paradigm Smart community 5.0 to replace Industrial Revolution 4.0, which dehumanized people and communities. Scientists have considered that the minimum effective dose of a probiotic is approximately 10^8–10^9 cells per mL despite there is no evidence-based consensus on the optimal concentration of bacteria pro-dose [17]. The fresh products, including probiotics, contain live starter culture bacteria, while the extended shelf-life products are in doubts of their content of live microorganisms. The most commonly used bacteria, which are found in various probiotic-dairy products, are Bifidobacteria and Lactobacilli. In addition to their desired health and clinical properties, probiotics must meet several basic requirements for the development of marketable probiotic products. The most important requirements are that probiotic bacteria must survive in sufficient numbers in the product, which their physical and genetic stability must be guaranteed during the storage of the product [18].

In recent years, the proposal of new probiotic-related concepts has gained attention. This is because it has been recognized that some mechanisms and clinical benefits cannot be directly related to the live microorganisms [19]. To survive in milk, they need some supplements that act as a growth promoter. Fermented soy milks can be developed using some probiotic cultures as *L. acidophilus, L. casei* as well as *Bifidobacterium pseudolongum* [14, 16, 20], due to the presence of oligosaccharides, belong to non-digestible starch, as Raffinose and Stachyose [21, 22]. These are prebiotic substances (**Figure 2**), which support the growth of probiotic microorganisms (growth promoter). Prebiotics include fructans, oligosaccharides, arabinooligosaccharides, isomaltooligosaccharides, xylooligosaccharides, resistant starch, lactosucrose, lactobionic acid, galactomannan, psyllium, polyphenols, and

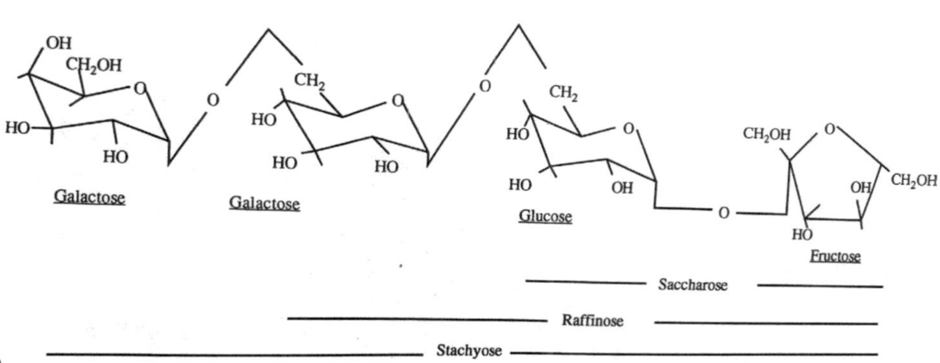

Figure 2.
Structure of Raffinose and Stachyose of soybean.

polyunsaturated fatty acids [23]. Soy flour contains fiber as much as 12 g/ 100 g products, equivalent to 1/5 of 60–80 g/ day recommended to maintain a large bowel flora of 10^{14} microorganisms, typically for a healthy human colon.

Lactobacillus casei and especially Bifidobacterium are among the probiotic cultures that defined by Fuller as a live microbial feed supplement which beneficially affect the host (animal) by improving its intestinal microbial balance. However, this definition of Fuller was improved by [24] as "One product contains specific live microbes at the sufficient numbers and change the microbial balance leading to health effect for the host." Probiotic's definition is living microorganisms, which when ingested in sufficient amounts, beneficially influence the health of the host by improving the composition of intestinal microflora [25]. At present, [2, 26] explained that the most frequently utilized species for promoting the health were:

1. Bifidobacterium: B. bifidum, B. longum, B. infantis, *B. breve*, B. adolescentis

2. Lactobacillus: L. acidophilus, L. casei, *L. rhamnosus*, *L. fermentum*, *L. reuteri*

3. Streptococcus: *S. faecium*

4. Propionibacterium: P. freudenreichii subsp. Shermanii

5. Saccharomyces: Saccharomyces boulardii

Many probiotic products on the market contain Lactic Acid Bacteria, the population of which tend to be a relatively small proportion of the total gut microflora. There are more than 400 bacterial species in human flora, but only ten species cohabit at the highest population levels. *Bifidobacterium* sp., is the most popular probiotic [14]. These are strict anaerobic and most of them are very sensitive to contact with atmospheric oxygen. The most abundant population belong to the gram-negative Bacteriodes genus (30% of the dominant human flora) and to the Gram-positive Eubacteria, Bifidobacterium, Ruminococcus and the different Clostridia species. Bifidobacteria (**Figure 3**), are accounting for approximately 1% of total microflora.

Bifidobacteries as a single culture cannot grow easily in dairy milk without supplementations despite it grows well in soymilk [14]. The yoghurt bacteria *L. delbrueckii subsp. bulgaricus* and *S. thermophilus* are not mentioned as probiotic culture (false claimed) because they are not typical forming the intestinal flora of man. Despite it, yoghurt that made from dairy milk or yoghurt-like product from vegetable extract are considered as the best-known food vehicle for probiotics because, beyond its own physicochemical and functional characteristics [2, 14].

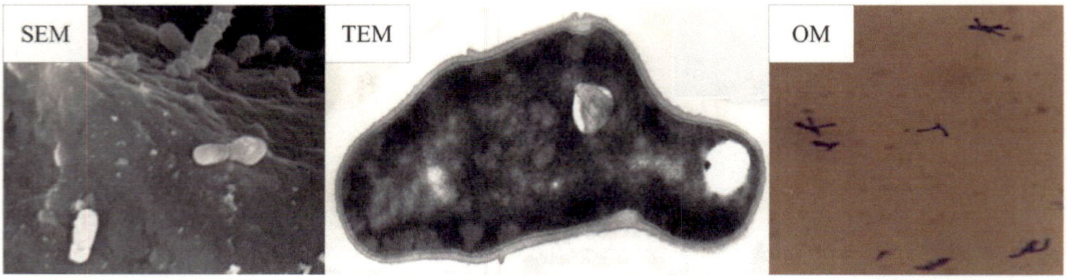

Figure 3.
Bifidobacterium pseudolongum viewed at scanning electron microscopy (SEM), transmission electron microscopy (TEM) and ordinary microscope (OM).

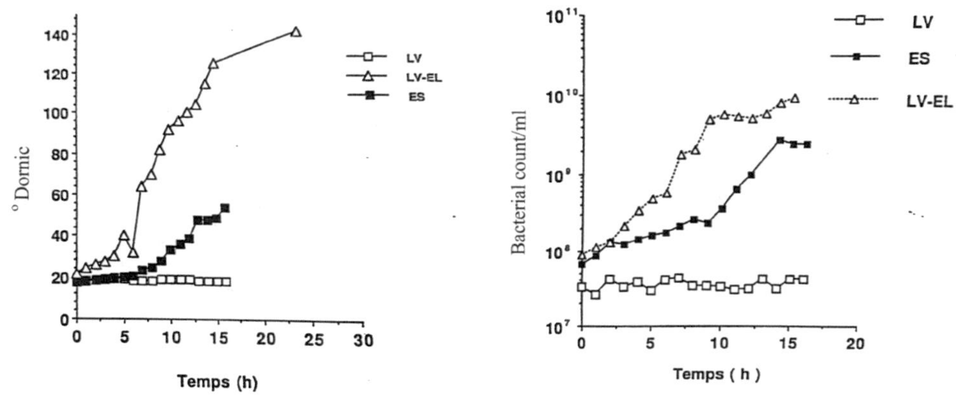

Figure 4.
Acidity (left) and bacterial counts (right) of Bifidobacterium sp. in yoghurt (CM), CM+ yeast extract and Soyoghurt (SE).

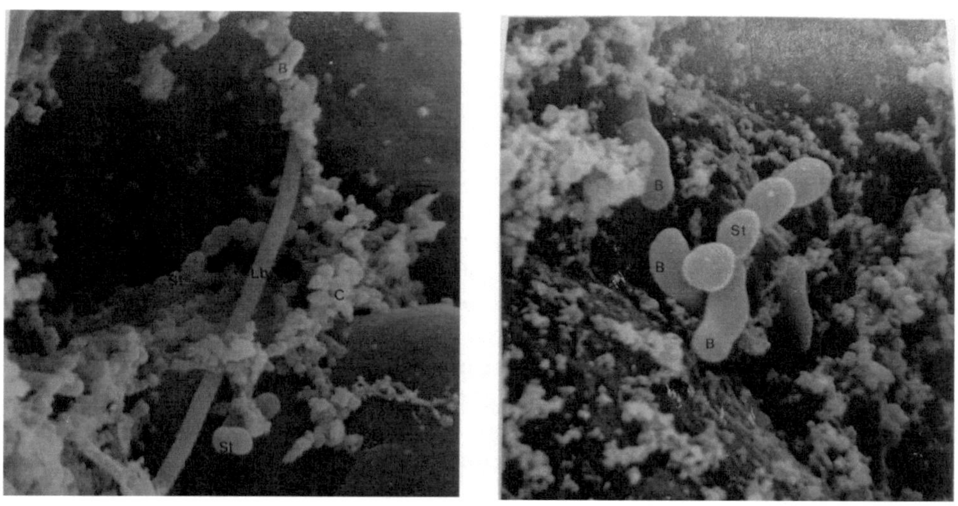

Figure 5.
B. pseudolongum in yoghurt (left) or soyoghurt (right).

Role of probiotic cultures to give positive impact for human health depends on if they are life or not in media, as in **Figures 4** and **5** above [14]. These probiotics microbes require a great number of bacteria that genetically stable since the manufacturing, during storage and expressing their health benefit after consumption especially during transit gastric. Consumers will accept the probiotic products that have a pleasant aroma and taste. It should bear in mind that the development of these bacteria either indirectly to fermented milk or directly to human gut also depends on the type of food ingested by humans. Human, included infant live under rather unnatural conditions lead to unfavorable condition causing illness. We eat a great deal of processed antibacterial substances ranging from vinegar to antibiotics, and in many cases sterile food which many affects our access to and colonization by certain type of bacteria. The healthful effect of probiotic -containing fermented milk is presented supplemented or not by prebiotic substances are in (**Table 3**).

The food in Indonesia is quite different from those in Western countries or Japan. The predominant food ingested by the Indonesians are food from vegetable origins, while in Western countries are from animal origins (meat, eggs, and milk)

Heath benefit of Fermented Probiotic Milk	
• Intestinal microbial balance	• Antidiarrhea
• Improved absorbability of certain nutrients	• Anticancer effect
• Alleviation of lactose intolerance symptom	• Antagonist environment. For pathogens
• Metabolism of some drugs	• Blocking adhesion sites from pathogens
• Improvement of intestinal motility	• Inactivating enterotoxins
• Stimulation of immune system	• Alleviating constipation
• Serum cholesterol reduction	• Relieving vaginitis
• Resistant to enteric pathogen	• Small bowel bacterial overgrowth
[1, 26, 27].	

Table 3.
Health effect of fermented milk's probiotics.

and in Japan which are from sea food origins. Some of the health and nutritional benefit of probiotics cultures containing LAB or others for human are generally are:

6.1 Enhanced lactose digestion

Lactose, the main carbohydrate in mammalian milk is fermentable compound, and digested by the intestinal enzyme, lactase, present especially in infants and lactose persistent adults. However, in large percent of the world population (70%) who are lactose non-persistent, lactose escapes digestion in the small bowel and is fermented in the colon with the evolution of gases (CO_2, H_2, CH_4) and the production of osmotically simple sugars. The gas, especially H_2 will be Trans located to lungs and can be detected as hydrogen exhalations. Improvement of lactose digestion and avoidance of symptoms of intolerance in lactose mal-absorbers are the most profoundly studied health—relevant effect of fermented milk products, despite are not specifically probiotics effects. The word lactose intolerance is well accepted rather than lactose mal-absorbers, due to no absorption in the proper term of lactose in small intestine. The symptom released due to the absence of or insufficient amount of β-*galactosidase* (*lactase*) in the human small intestine to digest the milk sugar, lactose. The prevalence of primary lactose intolerance is 3–5% in Scandinavia, 17% in Finland and Northern France, 65% in southern France, 20–70% in Italy, 55% in Balkans region, 70–80% in Africa, 80% in Central Africa, 90–100% in Eastern Asia, 30% in Northern India, 70% in southern India and 80% in black American [28]. While secondary form of lactose intolerance may be due to inflammation or functional loss of the mucus of small intestine (enteritis, bacterial or parasitic infection or small bowel syndrome) and by protein-energy malnutrition. It is generally accepted that lactose intolerance symptoms will appear after consuming more than 250 mL a day of whole milk [2]. Fermented milk contains less lactose (**Table 4**).

Lactase deficiency is due to low activity of intestinal lactase (lactase phlorizine hydrolase's in enterocyte of villi). The symptoms associated with lactose intolerance included: bloating, belching, flatulence, cramping and watery stools. Individual with this condition experience gastric distress when they consume fresh, unfermented dairy products. Fermented milk products can especially improve lactose digestion in individual lactose intolerance (**Table 5**).

The explanation will be given that microbial β-*galactosidase* auto digest this lactose and reduce the content of lactose present in dairy products between 20–30% of original lactose [33]. Some developed fermented milk has reduced lactose content around 30% with the range from 14 to 74% in fermented milks. The reduction of lactose

Fermented milks	**Lactose contents (%), at**		**Reduction (%)**
	0 h	**8 h**	
1. Cow milk yoghurt	4.31.	3.72	13.7
1. Cow milk yoghurt	4.4	1.9	74
1. Goat milk yoghurt	4.55	3.6	19.5
1. Goat milk yoghurt	4.4	3.0	31
1. 4+ Soy Growth Factors	4.9	3.85	21.4
1. Bifidus cow milk	4.2	3.28	21.9
1. Bifidus cow milk	4.7	3.5	30
1. 6+ Soy GF	4.7	4	14.8
1. 1+ 8	4.31	1.9	56
1. Bifidus goat milk	4.2	2.76	34.3
1. Acidophilus cow milk[b]	5	3	40
1. Cow milk[b]	5	3	40
1. Lactococci fermented milk[c]	4.5	3.4	24.5
1. Citrat-Lactococci fermented milk[c]	5.5	3.65	33.4
1. Leuconostoc fermented milk[c]	6.5	3.68	43.3
1. 13 + Bifidobacteria	3.4	2.58–3.29	> 25
1. 14 + Bifidobacteria	3.65	2.51–3.55	> 32
1. 15 + Bifidobacteria	3.68	2.92–3.43	21

[a]Sugars were detected by HPLC using Column Aminex HPX-87 H at flow rate 0.4 ml/min and mobile phase acetonitrile 30% in 0.005 M H2SO4 (Bio-rad), temp 65 C [2, 28–31].

Table 4.
Reduction of lactose content of some fermented milks[a].

Fermented milks	**% lactose**	**No cell/ g prod exceptb.**	**Lactase (mg/h/g prod)**
Yoghurt	4	3×10^8	0.64
Pasteurized yoghurt	4	3.4×10^6	0.07
Cultured milks	4.3	2.8×10^6	0.02
Sweet acidophilus milk	4.8	1.1×10^7	0
Pasteurized milk	4.9	Not measured	0
Bifidobacterium breve[b]	3.4	8.7 log CFU/mL	NA
Bifidobacterium longum[b]	3.65	8.7 log CFU/mL	NA
Bifidobacterium bifidum[b]	3.68	8.7 log CFU/mL	NA

[29, 32].

Table 5.
Percent lactose, cell counts and lactase activity of fermented milk fed to lactase deficient subjects.

varies either with raw milk, material, bacterial species used and its combination or the present of growth factors. It is therefore, Fermented milk reduce lactose content and consumers will adapt to different levels of lactose in any commercial fermented milks.

These still alive microbes will implant in the small intestine of humans and continue to digest of lactose lead to better adaptability to the present of such compound or giving improve to lactose tolerance. Yoghurt bacteria, *L. acidophilus*, and *Bifidobacterium* sp. are now well known can survive during gastric transit alive, numerous, and active in the human intestine despite it is not natural inhabitants. These bacteria then deliver their lactose-metabolizing enzyme to its site of action over an extended period. Therefore, probiotic-containing fermented milk could partially explain the better lactose digestion after ingestion of milk.

6.2 Colonization and role anti-pathogenic bacteria

The antimicrobial or antagonistic activity of probiotic cultures is an important property that people need to understand for using it as health benefit microbes for human beings. Breast-fed infants thrive better than bottle-fed or weaned babies do. This, in part can be associated with a predominant bifid gut flora that inhibits growth of coliforms, enterococci, and clostridia [34]. *Bifidobacterium* sp. account for 92% of the intestinal flora of breastfed infants, but only 20% of bottle-fed or weaned infants. The production of acetic and lactic acid has caused lower pH to inhibit pathogenic bacteria lead for children with high numbers of bifidobacteria to resist some enteric infection very effectively. Regularly supplementation of the infant diet with bifidobacteria can maintain normal intestinal condition. When *B. longum* administered orally to germ-free mice, the bacteria colonize the intestinal tract and reach a concentration of 10^{9}–10^{10}/g intestinal content in 2–3 days. Ability to transit gastric of living probiotic bacteria is important criteria to be used as human health promoting bacteria [35–37]. Translocation of the colonized *B. longum* to the mesenteric lymph node (MLN), liver and kidney occurs between 1–2 weeks after the association caused neither infection nor harmful effect. When *E. coli* O: 111 or O: 157 was administered orally to germ-free mice, they will be translocated to various organs occurred and the mice died by endotoxin shock or organ failure. When *B. longum*—mono associated mice challenged *E. coli* O: 157, the intestinal count of strain O: 157 suppressed at a low concentration and no death after 5 weeks. When with *s* O: 111 at a lethal dose, death was avoided [38, 39]. Treatment and prevention of infectious diarrhea are probably the most widely accepted health benefits of probiotic microorganisms. Rotavirus is the most common cause of acute infantile diarrhea in the world. Well-controlled clinical studies have shown that probiotics such as *L. rhamnosus GG, L. reuteri, L. casei, strain Shirota*, and *B. animalis Bb12* can shorten the duration of acute rotavirus diarrhea with the strongest evidence pointing to the effectiveness of *L. rhamnosus* GG and *B. animalis* Bb12 [40].

6.3 Role anti-cancer, anti-hypercholesteremic and anti-virus

In order to permanently establish the bacterial strain in the host's intestine, the organism must be able to attach to intestinal mucosal cells. The intestinal microflora within a given individual are remarkably stable, although major differences may exist among different persons.

Nevertheless, administration of especially probiotic cultures to either newborn or adult results in certain change in the microbial profiles and metabolic activities of feces. The Fermented Milks and Lactic Acid Beverages Association in Japan have introduced a standard of a minimum of $>1 \times 10^{7}$ CFU/ml or CFU/g viable probiotic cells for fresh dairy products [38]. It has suggested a minimum viable number of 10^{6} CFU/ml or grams but recommended 10^{8} CFU/g to compensate the reduction through passage through the gut.

In contrary to the general definition of probiotics, dead probiotics or bacterial' cell components can still combat cancer effectively [41]. In fact, it has been proposed that inactivated or non-viable probiotic cells, as well as their metabolic byproducts (herein referred as "postbiotics" and "paraprobiotics"), also have the ability to provide benefits to the host's health [19].

Probiotics are able to conjugate bile acids and to assimilate cholesterol. The feeding of milk formula supplemented with L. acidophilus to infants was shown to result in lower level of serum cholesterol [42]. While culture of *S. thermophilus* assimilated less cholesterol than *L. delbrueckii subs. Bulgaricus*. L. acidophilus assimilated more cholesterol than those S. thermophiles and a commercial yoghurt. Bifidobacteria were also actively assimilated cholesterol. [43] have studied the influence of probiotic yoghurt on serum lipids in women, found that the concentration of total cholesterol, LDL Cholesterol and triacyl glycerol on serum decreased after consuming the standard and this reduction did not depend on the cholesteroemi status (**Table 5**).

However, the beneficial impact of probiotic is not limited to the gut-associated diseases alone, but also in different acute and chronic infectious diseases. This is because probiotics are able to modify the intestinal microbial ecosystem, enhance the gut barrier function, provide competitive adherence to the mucosa and epithelium, produce antimicrobial substances, and modulate the immune activity by enhancing the innate and adaptive immune response. Dairy-based matrix is suitable for proliferation of probiotics due to supplying a rich source of carbon and essential amino acids as a consequence of lactose hydrolyzing and proteolytic systems involved casein utilization [44]. Oral administration of *L. casei* SY13 also significantly enhanced the gut microbial diversity [45]. When, such change even minor applied to pathologic situation, they are sufficient to beneficially alter the course of disease, especially that are associated with the developments of colon cancer [46] as in **Table 6.**

It has explained that some research results using *L. acidophilus* (1×10^9 CFU/day) and *B. bifidum* (1×10^{10} CFU/day) during 3 weeks reduced fecal activity of

	Probiotic	**Experiment**	**Major finding**	**Reference**
in vivo	Yoghurt (unknown)	Human	Reduced total cholesterol and LDL	Agerhol Larsen *et al*. (2000)
	Fortified buffalo	Rat	Reduced total cholesterol, LDL-cholesterol, and triglyceride	Abd El-Gawad *et al.* (2005)
	L. plantarum	Mice	Reduced blood cholesterol Decreased triglycerides	Nguyen *et al.* (2007)
	*L. acidophilus*145, *B. longum* 913	29 woman/21 week	Up HDL-cholesterol, down LDL/HDL ratio	Khani *et al*., 2012
in vitro	*L. fermentum*	Culture media	BSH activity	Pereira *et al.* (2003)
	L. acidophilus *L. bulgaricus* *L. casei*	Culture media	Assimilation of cholesterol Attachment of cholesterol onto cell surface Disrupt the formation of cholesterol micelle Deconjugation of bile salt Exhibited bile salt hydrolase activity	Lye *et al*. (2010)
	Lactobacillus plantarum CGMCC,	Colorectal cancer patients	Probiotics decreased the serum zonulin concentration,	Liu *et al*., 2013
	L. rhamnosus GG, *B. lactis* and inulin	37 colon cancer/12 week	Down colorectal proliferation & necrosis	Khani *et al.,* 2012

[47–49].

Table 6.
Beneficial role of probiotics cultures.

nitro reductase, *azoreductase, and* β-*glucoronidase*, which is considered implicated to cancer in colon activity after daily intake of fermented vegetables for several weeks [50]. The anti-carcinogenic effect of bifidobacteria may be the results of direct removal of pro-cancerigenes, indirect or activation of the body's immune system [34]. By directly removing pro-cancerigenes, bifidobacteria, indirectly, *B. longum* remove the source of pro-carcinogenes or the enzyme, which lead to the formation of liver tumor, developed in mice when an intestinal flora of *E. coli, Enterococcus faecalis* and *C. paraputrificum* was presence.

Probiotics mainly colonize in the gut, but it has a fundamental impact on the systemic immune response (Immune enhancer), and exert the immune responses at distance ucosal site, including the lung [51]. There is accumulating evidence that bidirectional communications exist between gut and lung, which is called the gut-lung axis, in which probiotics modulate mucosal immune function. *L. acidophillus* and *L. casei* have some positive role to inhibit viral replication [52]. This bidirectional crosstalk is involved in the support of immune homeostasis. It is believed that the gastrointestinal inflammation results in lung inflammation through this connection [23].

It was found that COVID-19 patients with GI symptoms such as diarrhea experienced more severe respiratory disorders than those without GI symptoms. Probiotic consumption improves the level of type I interferons, antigen-presenting cells, NK cells, and T and B cells in the lungs' immune system. Probiotic administration can also improve the pro- and anti-inflammatory cytokines, helping to clear the viral infection by minimizing the cell damage in the lungs [51]. Antunes et al. [53] explained that some research about the role of fermented milk to fight COVID-19. They have been conducting in US as for the moment, a clinical trial is being performed on the effect of *L. rhamnosus* GG on the microbiome of household contacts exposed to COVID-19 (U.S. National Library of Medicine, Clinical Trial gov Identifier NCT04399252, 2020).

Other ongoing studies using probiotics to prevent or treat COVID-19 are "Evaluation of the probiotic *Lactobacillus coryniformis* K8 on COVID-19 prevention in healthcare workers" (U.S. National Library of Medicine, Clinical Trial gov Identifier NCT04366180, 2020).

6.4 Development of flavor and enhance consumers' acceptance

The wholesome and goodness of food products included the flavor influence acceptability of consumers. While, Muslim's community, they also need halal status of food and beverage products. Flavor is defined as complexes' sensation of taste, odor, and some physical aspects as texture, and hardness. Odor relates the volatile components that released during mastication or as the results of bacterial activity [2].

Some odors have reduced the perception score of consumers lead to rejection from people, as beany flavor of soybean foods. Fermented milk can mask this beany flavor [14, 20, 54, 55] have studied the odor of some fermented milk (**Tables** 7 and **8**) lead to more acceptability in diverse consumers.

Fermentation ability to produce differ flavor is depended also on genotypes of dairy animals. The genetic polymorphisms were related to the milk titrable acidity, alcohol stability, pH and bacterial count, as a possible effect of CN and LGB gen polymorphisms [56]. It also influenced on the physical properties of cultured milk as for example cheese syneresis, particle size distribution and odor production [57].

Preliminary data of our research on milk polymorphisms of Saanen Goat and Ettawa grade goat in Indonesia explained that there were at least 9 alleles of CSN1S1: AE (7.01%), AF (15,79%), AN (5,26%), A01 (1,75%), EE (8,77%), EF (47,37%), FF (10.53%), F01 (1,75%), NN (1,75%) for Saanen goats and 5 alleles of CSN1S1: AA

Species	Acetaldehyde	Acetone	Butanal	Butanol 2	Ethanol
L. delbrueckii subsp. Bulgaricus	3,1	1.4	0.1	0.17	0.5
L. helveticus	3.95	3.05	0.1	0.55	0.2
L. acidophilus	5.75	3.3	0.55	0.6	0.5
L. casei	3.2	4	0.1	0.8	0.4
L. casei strain Shirota	4.81	19.5	—	—	232
Average Lactobacillus	4.16	6.25	0.19	0.42	46.6

**Skim milk, [54].*

Table 7.
*These bacteria have been developed in skim milk. * at Lactobacillus (Lactobacillus*).*

Compounds	Yakult	Yoghurt 1	Kefir
Acetaldehyde	4.81 ± 4,3^{a}	14,82 ± 1,99^{b}	4.48 ± 1,41^{a}
Acetone	19.58 ± 13,7^{d}	39,92 ± 4.56^{e}	7,51 ± 3,38^{c}
Ethanol	232,1 ± 158^{g}	1,56 ± 1,32^{f}	214 ± 63^{g}
Diacetyle	14,63 ± 3,8^{i}	2,72 ± 1,49^{h}	80 ± 55^{j}

** $^{a-j}$ means the different letter in the same row was significantly different [55].*

Table 8.
Odor of fermented milk containing lactic acid bacteria cultures (ppm).*

(67,65%), AE (2,94%), AF (14,71%), AN (11,76%), EF (2,94%) of Ettawa grade goat (Murti, 2020; unpublished data). These polymorphisms can influence physico-chemical and flavor of fermented milk.

7. Conclusion

Fermented milks especially which use intestinal microflora, known as probiotic cultures link to many aspects of host health. Dairy-based matrix is suitable for proliferation of probiotics due to supplying a rich source of carbon and essential amino acids as a consequence of lactose hydrolyzing and proteolytic systems involved casein utilization. However, we still have limited understanding within the mechanisms involved in these systems, particularly with regard to colonization resistance. Despite a lot of research to date, many questions still remain as to how the modulations of the microflora are able to offer benefits for the host in either good health or illness. Knowledge of its composition and activities are central to this, as is a better understanding of how influence the toleration of microflora. Colonization of the gut may not be an absolute necessity since the healthy host influence the toleration of the strain becoming tolerant, thereby removing its immunological capacity to be a probiotic. Since different probiotic strains are known to have different immunological and microbiological properties, molecular studies should enable specific components of bacterial phenotype, and so identify strains that can benefit the host with diarrhea or inflammatory diseases as well as their role to fight Covid-19.

Author details

Tridjoko Wisnu Murti
Laboratory of Dairy Science and Milk Industries, Faculty of Animal Science, Universitas Gadjah Mada, Yogyakarta, Indonesia

*Address all correspondence to: tridjokomurti@mail.ugm.ac.id

References

[1] Danone. Les produits laitiers fermentes protegent-ils contre le cancer colorectal? News Letter no 26. 2003[b]

[2] Murti, T.W. The Health Benefit Role of Fermented Milk. In: Proceeding of the Nat. Sem., Fac. of Animal Sci. on Food Safety on Animal Products.; 14 november 2005; Universitas Gadjah Mada at Yogyakarta, Indonesia; 2005. p. 28-44

[3] Murti, T.W., E. Robiyati, H.L. Jundi, F.Ramadhani, B. Rustamaji & Y.Y. Suranindyah. Development of Fermented Mare's milk Using Mixed probiotic Cultures. Media Peternakan (TASJ). 2016:39:9-13. DOI: 10.5398/medpet.2016.39.1.9

[4] Mayo, B., M. S. Ammor, S. Delgado, and Á. Alegría. Fermented Milk Products. In: Tamang J.P., Kailasapathy K, Editors. Fermented foods and beverages of the world. Boca Raton, FL USA: CRC press. 2010.p. 263-288. DOI: 10.1201/EBK1420094954-c9

[5] Murti, T. W. *Post-Harvest Treatment of Milk.* Gadjah Mada University Press. 2016. 261 p.Yogyakarta, Indonesia.

[6] Banerjee, S., R. Pandey, T. Gorai, S.L. Shrivastava, and S. Haldar. Review on Soy Milk and Other Soy Milk Based Products. Int. Res. J. Food Nutr. 2019:1 (1): 1-5. Available from: https://scirange.com/pdf/28-irjfn-2018.pdf [Accessed: 2020-10-26]

[7] Danone. The intestinal microflora. Understanding the symbiosis. Nutrition and health collection. John Libbey Eurotext, Paris. 2003[a]

[8] Ranadheera C. S., C. A. Evans, S. K. Baines, C. F. Balthazar, A. G. Cruz , E. A. Esmerino, M. Q. Freitas. Probiotics in Goat Milk Products: Delivery Capacity and Ability to Improve Sensory Attributes. Comprehensive. Rev. Food Sci. and Food Safe. 2019:18:867-882. DOI: 10.1111/1541-4337.12447

[9] Faccia, M., A. G. D'Alessandro, A. Summer and Y. Hailu. Milk Products from Minor Dairy Species: A Review. Animals. 2020:10:1-25. DOI: 10.3390/ani10081260

[10] Murti, T.W. A study on flavor and various of organic acid of fermented milk goat by using *Lactobacillus casei. J. Indonesia. Trop. Animal Agric.* 2007:32 (2):230-235. Available from: http://www.jppt.undip.ac.id/pdf/32(4)2007p230-235.pdf [Accessed: 2020-12-22]

[11] Murti, T.W., B.T. Santoso and A. Latif. The Quality of Low Fat-Fermented Goat Milk and Cow Milk Containing Probiotic Cultures. IOP Conf. Ser: Earth Environ. Sci. 387:012132. 2019. Available from https://iopscience.iop.org/article/10.1088/17551315/ 387/1/012132/pdf [Accessed: 2020-12-22]

[12] Bernat, N., M. Chafer, A. Chiralta, and C. Gonzalez-Mart. Vegetable milks and their fermented derivative products. Int. J. Food Stud. 2014:3:93-124. DOI: 10.7455/ijfs/3.1.2014.a9

[13] Mazumder, A. R, and A. R. Begum. Soymilk as source of nutrient for malnourished population of developing countries: A review. Int J. Adv. Sci and Techn. Rech. 5: 192-203. 2016. Available from: www.rspublication.com/ijst/index.html [Accessed: 2020-10-26]

[14] Murti, T. W. *Growth, Sensory, and Biochemical Effects of Fermented Soy Milk using Lactic Acid Bacteria and Bifidobacteria, as Compared to Those of Fermented Cow Milk. Dissertation of Caen University, France*; 1993.

[15] Ranadheera C. S.,, J. K. Vidanarachchi, R. S. Rocha, A. G. Cruz, and S. Ajlouni. Review: Probiotic

Delivery through Fermentation: Dairy vs. Non-Dairy Beverages. Fermentation 3. 2017:67:1-67. DOI: 10.3390/fermentation3040067

[16] Murti, T. W., S. Roger, C. Bouillanne, M. Landon, and M. J. Desmazeaud. Growth of Bifidobacterium sp CNRZ 1494 in Soy Extract and Cow Milk. Effects on Aroma Compounds. Sci. Aliments. 1992:12: 429-439. Available from: https://www.researchgate.net/publication/341866572_Croissance_de_Bifidobacterium_SP_CNRZ_1494_dans_l'extrait_de_soja_et_le_lait_de_vache_Effets_sur_des_composes_aromatiques [Accessed: 2020-10-26]

[17] De Cerbo A. and B. Palmieri. Review the market of probiotics. Pak. J. Pharm. Sci., 28: 2199-2206. 2010. Available from: www.pubmed.ncbi.nlm.nih.gov/26639512/ [Accessed: 2020-10-20]

[18] Heller, K.J. Probiotic bacteria in fermented foods. Products characteristics and starter organisms. Am J. Clin Nutr. 2001:73(S):374-379 S. DOI: 10.1093/ajcn/73.2.374s.

[19] Vallejo-Cordoba, B., C. Castro-López, H. S. García, Aa. F. González-Córdova, and Adrián Hernández-Mendoza. Postbiotics and paraprobiotics: A review of current evidence and emerging trends. In: ADVANCES IN FOOD AND NUTRITION RESEARCH Probiotic and Prebiotics in Foods: Challenges, Innovations and Advances. Eds: A. G. Da Cruz, E. S. Prudencio, E. A. Esmerino, And M. C. Da Silva. Academic press, Cambridge, MA-USA. 2020:94:91-114. DOI: 10.1016/bs.afnr.2020.06.001

[20] Murti, T. W., G. Lamberet, C. Bouillanne, M. J. Desmazeaud and M. Landon. Lactobacilli growth in Soy-Milk. Effect on Viscosity, Volatile compounds and Proteolysis. Sci. Aliments. 1993.13: 491-500. Available from: https://agris.fao.org/agris-search/search.do?recordID=FR9305653 [Accessed: 2020-12-08]

[21] Bengmark, S. Lactic Acid Bacteria and Plant Fibers. Treatment in Acute and Chronic Human Disease. In Prebiotics and Probiotics Ingredients. Eds. S. S. Cho and E. T. Finocchiaro. CRC Taylor and Francis, Baca Raton, FL- USA. 2010. Available from: hwww.researchgate.net/publication/329528682_Lactic_acid_bacteria_and_plant_fibers_Treatment_in_acute_and_chronic_human_disease [Accessed: 2020-11-29]

[22] Cho, S. S., and E. T. Finocchiaro. Natural resistant starches as prebiotics and synbiotics. In Prebiotics and Probiotics Ingredients. Eds. S. S. Cho and E. T. Finocchiaro. CRC Taylor and Francis, Baca Raton, FL- USA. 2010. Available from: www.researchgate.net/publication/329541982_Natural_resistant_starches_as_prebiotics_and_synbiotics [Accessed: 2020-12-08]

[23] Olaimat A. N., I. Aolymat , M.Al-Holy, M. Ayyash, M. A. Ghoush, A. A. Al-Nabulsi, T Osaili, V. Apostolopoulos, S. Q Liu and N. P. Shah. The potential application of probiotics and prebiotics for the prevention and treatment of COVID-19. npj Science of Food. 2020:4(17). DOI: 10.1038/s41538-020-00078-9

[24] Schrezenmeir, J. and M. de Vresse. Probiotics, prebiotics and synbiotics, approaching definitions. Am J. Clin. Nutr. 2001:73(S): 361-364. DOI: 10.1093/ajcn/73.2.361s.

[25] Abou-Zeid, N. Review: Probiotic Bacteria in Dairy Products. IJPAS 2, Issue-12,37-55. 2015. Available from: https://www.academia.edu/38566483/Review_Probiotic_Bacteria_In_Dairy_Products [Accessed: 2020-05-22]

[26] Lee, Y.K. Probiotic Microorganisms. In. Handbook of Probiotics and

Prebiotics. 2nd ed. eds. Lee, Y. K., and S. Salminen John Wiley and Son Publ. New Jersey, USA. 2009. DOI: 10.1002/9780470432624

[27] Collado M. C. Role of Probiotics in Health and Diseases in. Handbook of Probiotics and Prebiotics. 2nd ed. eds.in. Lee, Y. K., and S. Salminen. John Wiley and Son Publ. New Jersey, USA. 2009.p. 257-376. DOI: 10.1002/9780470432624.ch4

[28] Alm, L. Effect of Fermentation on Lactose, Glucose, and Galactose Content in Milk and Suitability of Fermented Milk Products for Lactose Intolerance Individuals. J. Dairy Sci. 1982: 65: 346-352. DOI: 10.3168/jds.S0022-0302(82)82198-X

[29] Barona, M., D. Roy, and J.C. Vuillemard. Biochemical characteristic of fermented milk produced by mixed cultures of lactic starters and bifidobacteria. Lait. 2000:80:465-478. DOI: 10.1051/lait:2000138

[30] Murti, T.W. Biochemical change of carbohydrate, organic acids, and volatile key compounds of bifidus milk. 52 the International symposium of EAAP-The Hague-Netherland; 2000.

[31] Murti, T.W. Consumer's acceptability and biochemical change of carbohydrate, and organic acids of developed yakult and kefir. Int Conference on Functional Food, Faculty of Agricultural Technology, Gadjah Mada University. Yogyakarta; 2003.

[32] De Vresse, M., A. Steplemann, B. Richter, S. Feaselau, C. Lauc, and J. Schrezemeir. Probiotics. Compensation for lactase insufficiency. Am J. Clin Nutr. 2001:73: 421-429S. DOI: 10.1093/ajcn/73.2.421s

[33] Ershidat, O.T.M., and A.S. Mazahreh, Probiotics bacteria in fermented dairy products. Pakistan J. Nutr. 2009:8:1107-1113. DOI: 10.3923/pjn.2009.1107.1113

[34] Hughes, D.B., and D.G.Hoover. Bifidobacteria: Their potential for use in American dairy products. Food Technol 45: 74-83. 1991. Available from: www.semanticscholar.org/paper/ BIFIDOBACTERIA-%3A-their-potential-for-use-in-dairy-HughesHoover/40d059355da9438a534b19fe178bae74a21f0797#citing-papers [Accessed: 2005-02-27]

[35] Berrada, N., J. F. Lameland, G. Laroche, P. Thouvenot. Bifidobacterium from fermented milk: Survival during gastric transit. J .Dairy Sci. 1991:74: 409-413. DOI: 10.3168/jds.S0022-0302(91)78183-6

[36] Marteau, P. , P. Pochard, B. Fluorie, P. Pellier, L. Santos, S.F. Desjeux, and J.C. Rambaud. Effect of chronic ingestion of a fermented dairy products containing Lactobacillus acidophilus and Bifidobacterium bifidum on metabolic activities of the colonic flora in humans. Am. J. Clin Nut. 1990:52: 685-688. DOI: 10.1093/ajcn/52.4.685.

[37] Marteau, P., M. Minekus, P. Havenaar, and JHJ Hui in't veld. Survival of lactic acid bacteria in a dynamic model of the stomach and small intestine. Validation and the effect of bile. J. Dairy Sci. 1997:80: 1031-1037. DOI: 10.1093/ajcn/52.4.685.

[38] Ishibashi, N., and Ishimura. Bifidobacteria: research and development in Japan. Food Tech. 1993:47(6):129-130. Available from: https://www.researchgate.net/publication/312618753_Bidobacteria_Research_and_development_in_Japan [Accessed: 2005-02-27].

[39] Ishibashi, N., and S.Yamazaki. Probiotics and safety. Am J. Clin Nutr. 2001:73 (S):465-470. DOI: 10.1093/ajcn/73.2.465s

[40] Isolauri, E., P. V. Kirjavainen, and S. Salminen. Probiotics: a role in the treatment of intestinal infection and in

Inflammation. Gut. 2002:50(3):54-59. DOI: 10.1136/gut.50.suppl_3.iii54

[41] Bedada T. L., T. K. Feto, K. S. Awoke, A. D. Garedew, F. T. Yifat, and D. J. Birri. Review Probiotics for cancer alternative prevention and treatment. Biomedicine & Pharmacotherapy 2020:129:110409. DOI: 10.1016/j.biopha.2020.110409.

[42] Rasic, J.L.,. I.F.Vujific, M.Skrinjar, and M.Vulic. Assimilation of cholesterol by some cultures of lactic acid bacteria and Bifidobacteria. Biotechnol. Lett. 1992:14: 39-44. DOI: 10.1007/BF01030911

[43] Sadrzadeh-Yeganeh, H., I.Elmadfa, A.Djazayery, M.Jalali, R.Heshmat, and M.Chamary. The effects of probiotic and conventional yoghurt on lipid profile in women. Br.J.Nutr. 2010.103(12):1778-1783. DOI: 10.1017/S0007114509993801

[44] Khoshirdian. M., M. Yousefi, and A.M. ortazavian. Fermented Milk: The Most Popular Probiotic Food Carrier. In: ADVANCES IN FOOD AND NUTRITIONvRESEARCH Probiotic and Prebiotics in Foods: Challenges, Innovations and Advances. Eds: A. G. Da Cruz, E. S. Prudencio, E. A. Esmerino, And M. C. Da Silva. Academic press, Cambridge, MA-USA. .2020:94:1-34. DOI: 10.1016/bs.afnr.2020.06.007

[45] Yue Y, X. Xu, B. Yang, J. Lu, S. Zhang, L. Liu, K. Nassar, C. Zhang, M. Zhang , 1 X. Pang , and J. Lyv . Stable Colonization of Orally Administered Lactobacillus casei SY13 Alters the Gut Microbiota. BioMed Research International. 2020:2020:5281639. DOI: 10.1155/2020/5281639

[46] Bezkorovainy, A. Probiotics. Determinant of survival and growth in the gut. Am J. Clin Nutr. 2001:73(S): 399-405S. DOI: 10.1093/ajcn/73.2.399s

[47] Shi L. H., K. Balakrishnan, K. Thiagarajah, N. I M. Ismail and O.S Yin. Beneficial Properties of Probiotic. Tropical Life Sciences Research, 2016:27(2):73-90. DOI: 10.21315/tlsr2016.27.2.6

[48] Nazir, Y., S. A. Hussain, A. A. Hamid, and Y. Song. *Review Article* Probiotics and Their Potential Preventive and Therapeutic Role for Cancer, High Serum Cholesterol, and Allergic and HIV Diseases. Hindawi BioMed Research International. 2018:3428437: 1-17. DOI: 10.1155/2018/3428437

[49] Khani S., H. M. Hosseini, M. Taheri, M. R. Nourani and. A. I. Fooladi. Probiotics as an Alternative Strategy for Prevention and Treatment of Human Diseases: A Review. Inflammation & Allergy - Drug Targets, 2012:79-89. DOI: 10.2174/187152812800392832

[50] Wollowski, I., G. Rechkemmer and B.L. Pool-Zobel. Protective role of probiotics and prebiotics in colon cancer. Am J. Clin. Nutr. 2001:73 (2):451-455. DOI: doi: 10.1093/ajcn/73.2.451s.

[51] Sundararaman, A., M. Ray, P. V. Ravindra, and P. M. Halami. Role of probiotics to combat viral infections with emphasis on COVID-19. Appl Microbiol Biotechnol. 2020:104:8089-8104. DOI: 10.1007/s00253-020-10832-4

[52] Lehtoranta, L, S. Latvala and M. J. Lehtinen. Role of Probiotics in Stimulating the Immune System in Viral Respiratory Tract Infections: A Narrative Review. Nutrients. 2020. 12(10):3163. DOI: 10.3390/nu12103163

[53] Antunes, A. E.C., G. Vinderola, D. Xavier-Santosa, and K. Sivieri. Potential contribution of beneficial microbes to face the COVID-19 pandemic. Food Res. Int. 2020: 136. DOI: 10.1016/j.foodres.2020.109577

[54] Murti, T.W. Biochemical measures of aroma profiles and analysis of

consumer's acceptability of developed yoghurt using goat and cow milks. Report of research for Torray Sci Foundation; 1996.

[55] Murti, T.W. Development of an accepted fermented milks using health-promoting microflora for Indonesian consumers. Report of research for URGE New Dr batch I ; 1998.

[56] Kyselová, J., K. Ječmínková, J. Matějíčková, O. Hanuš, T. Kott, M. Štípková, and M. Krejčová. Physiochemical characteristics and fermentation ability of milk from Czech Fleckvieh cows are related to genetic polymorphisms of -casein, –casein, and -lactoglobulin. Asian-Australasian Journal of Animal Sciences. 2019:32(1):14-22. DOI: 10.5713/ajas.17.0924

[57] Ketto, I.A., J.Øyaas, T.Ådnøy, A.Johansen, R.B Schüller, J. Narvhus, Siv B. Skeie. The influence of milk protein genetic polymorphism on the physical properties of cultured milk. Int. Dairy J. 2017:78:130-137. DOI: 10.1016/j.idairyj.2017.11.009

Chapter 6

Microbial Diversity of Traditionally Processed Cheese from Northeastern Region of Transylvania (Romania)

Éva György and Éva Laslo

Abstract

The composition and production technology of the cheese are extremely diverse. There are a wide variety of microbial species on their surface, with a much smaller number inside of the product. The microbiota of the cheese may be composed of beneficial microorganisms, spoilage and foodborne pathogens. Identification and characterization of the microorganisms present in these products are important nutrition, food safety and technological aspects. During our work we evaluated the prevalence of allochthonous bacteria and microscopic fungi in traditionally processed cheeses from northeastern region of Transylvania, with classical microbiological culture methods. Based on the results the microbiota of the analysed cheeses was highly diversified. The identified bacteria with the highest prevalence from different selective media, were as follows: *Escherichia coli*, *Enterococcus durans*, *Enterococcus faecalis*, *Shigella flexnerii*, *Proteus vulgaris*, *Stenotrophomonas maltophilia*, *Staphylococcus equorum subsp. equorum, Staphylococcus equorum subsp. linens, Halomonas alkaliphila, Kocuria rhizophila, Hafnia paralvei, Bacillus licheniformis* and *Klebsiella michiganensis.*

Keywords: microbial diversity, traditionally processed cheeses, starter culture, allochthonous bacteria

1. Introduction

Cheese belongs to dairy products representing an important part of human balance diet for hundreds of years. It contains the essential macronutrients and micronutrients. Different factors as environmental conditions, geographical region, processing technologies and the local customs determine the development and release of the huge variety of cheese. One prevention method of milk spoilage over the centuries has been the production of cheese, meanwhile the nutritional benefits are conserving [1].

Cheese is the result of a fermentation process driven by lactic acid bacteria. Different bacteria, yeast and molds constitute the complex microbial ecosystem of this product. Throughout the ripening, the microbes release proteinases and lipases that are involved in the protein and fat hydrolysis, resulting amino acids and fatty acids. These compounds represent important flavour precursors for cheeses [2].

The cheese microbiota is involved in the development of the appearance texture, flavour and aroma. Both starter culture organisms and allochthonous microorganisms are part of the specific microbial ecosystem.

The quality and the processing of the milk, the environmental conditions (temperature, humidity, salt), the applied technology influence the quantity and the diversity of the specific species of microorganisms. Also, this fact is affected by the microbial conditions in the manufactures [3].

Cheese can be considered a privileged product with beneficial or negative impact on health, influenced by the nutritional composition, salt and bioactive compound content. The negative effect is related to the presence of pathogenic bacteria such as *Listeria* and *Escherichia coli* [1].

2. Factors controlling microbial growth in cheeses

In various types of cheese, the growth and development of microbes are influenced by physical and chemical parameters (water activity, pH, ripening temperature, redox potential), chemical composition (salt content, presence of nitrates) and by the metabolites as organic acids and the protein like compounds released by starter cultures named bacteriocins [2, 4].

Water activity (a_w) of cheese range between 0.917–0.988. Vast majority of non-starter bacteria, spoilage and pathogens are favoured by this condition. The optimum a_w of starter bacteria is under these values. Osmophilic yeasts support a_w less than 0.60 [4]. The a_w is related to sodium chloride (NaCl) and the produced metabolites (lactate, acetate, and propionate) content. These compounds decrease the vapour pressure of water as the result of their dissolution in the moisture of the cheese. Higher the concentration of these metabolites, the reduced the a_w is. Without vacuum or sealed packaging, the moisture content is reduced, due to evaporation, causing a gradient in the a_w of the cheese. This gradient commonly is proportional with the cheese size and is lower in the external part. At low a_w the lag phase of bacterial growth is prolonged [2].

The salt content decreases the a_w value of the cheese and consequently is an antimicrobial agent. For example, for the brined-cured cheeses at the beginning of the ripening process, the salt concentration is much higher on the cheese surface than in the inside the block. For this reason, salt tolerant microorganisms can grow on the surface of this type of cheeses. Surface-ripened cheeses microbiota include *Brevibacterium linens* and *Debaryomyces hansenii* growing at 15% NaCl with a_w ~0.916 [4]. Law salt content was associated with the development of coryneforms, micrococci, and staphylococci. The growth of these microorganisms is supported by 10–15% NaCl. 10% NaCl doesn't affect the development of *P. camemberti*. 20% NaCl is tolerated by several strains of *P. roqueforti,* whereas *Geotrichum candidum* is susceptible to salt [2]. Concentration of 15 g/L NaCl is used in manufacturing of Protected Designation of Origin (PDO) Serpa cheese [5]. NaCl can have a negative impact on cheese, affecting the enzymatic and microbial activities. This lead to the change of biochemistry processes as lactose metabolism, hydrolysis of fats and casein, the flavour compound development and the cheese pH formation. Paracasein aggregation or hydration is influenced by the salt -, calcium content and pH. These determinate the water binding properties of the casein matrix, the textural, rheological and cooking aspects and the ability to synerese [6].

Most bacteria grow best at neutral pHs. In acidic conditions ($pH < 4$) only several bacteria are able to grow, for example *Lactobacillus* spp. Molds and yeasts prefer pHs lower than 4.5. Viability of bacteria is affected by metabolites. Various weak organic acids (as sorbic and propionic acid) are causing bacterial lysis. These compounds cross the bacterial cell membrane releasing protons inside the cell and leading to acidification. In cheese environment, the main occurring organic acids are acetic, lactic and propionic acids. Propionic acid exerts antimicrobial effect against molds [4]. The slight acidic range in curd (pH = 4.5–5.3) contributes significantly to the growth and development of bacteria in cheese [7]. Microbial growth is determined also, by the ripening temperature. The optimum temperatures for the frequently used lactic acid bacteria, belonging to mesophilic and thermophilic bacterial groups, is 35°C and 55°C respectively.

Maturation temperature should play a role in preventing the growth of undesirable microbes as secondary flora, spoilage and pathogenic microorganisms, thus avoiding the losses [4].

The food hygiene is taken very seriously in the dairy industry. The quality of dairy products depends on raw milk quality and the processing conditions. Before ripening the number of microorganisms showed higher values in artisanal soft cheese from Germany named kochkäse, made from raw milk in contrast with cheese from pasteurised milk. After maturation, the amount of microorganisms was identical in the two samples. There is a close correlation between the temperature and ripening time. The ripening time is decreased at 35°C with the rising microorganism population. The maturation temperature at 15°C favour the lactic acid bacteria growth [8].

Redox potential (E_{redox}) values in cheese is approximately equal with E_{redox} = −250 mV, during the maturation process is reduced, due to unknown mechanisms. The redox potential is different inside and outside of the cheese that specifies the microbiota.

It was shown that the cheese surface is characterized by aerobic bacteria namely *Brevibacterium* spp., *Bacillus* spp. and *Micrococcus* spp. The predominant microbes inside the cheese are anaerobic or facultative aerobic microorganisms [4].

For the prevention of bacterial gas formation by coliforms and *Clostridium tyrobutyricum*, different chemicals are added to milk. In case of cheeses as Gouda and Edam cheeses potassium or sodium nitrate is added in amount of 20 g/100 L raw milk). In the milk or curd, the xanthine oxidoreductase catalyses the formation of nitrites from nitrates, that inhibit the aforementioned progress. Also carcinogenic compound as nitrosamine may resulted from the reaction of nitrites with aromatic amino acids [2].

Environmental factors influence the growth and development of lactic acid bacteria in cheese. The environmental factors and the various amount of NaCl, glucose and lactose affect the growth and biofilm formation of bacterial strains originated from Serbian traditionally processed cheese (*Enterococcus hirae* KGPMF9, *Lactococcus lactis* subsp. *lactis* KGPMF23, *Lactobacillus fermentum* KGPMF29, *Streptococcus uberis* KGPMF2). The limiting values for the factors were 6.5% of NaCl and 4°C. Except for *E. hirae* KGPMF9. *S. uberis* KGPMF2 different concentration of the carbohydrates and temperature of 37°C induced the planktonic growth and biofilm formation of the analysed bacteria. In case of biofilm formation basic media in contrast to acidic media was limiting factor [9].

3. Starter bacteria

In cheese processing lactic acid bacteria are used as starter cultures, contributing to organoleptic characteristics and preservation of the product. These strains are added in a different form, grown in milk or mediums. Also, in France or Italy the whey incubated the day before use is applied as starter culture. Starters comes from the fact that these bacteria start the lactic acid production. These belong to mesophilic and thermophilic bacteria [2, 10]. Mesophilic bacteria include lactococci (*Lactococcus lactis* subsp. *lactis* and *Lactococcus lactis* subsp. *cremoris*,) and often leuconostocs (*Leuconostoc mesenteroides* subsp. *cremoris* and *Leuconostoc lactis*). Thermophilic cultures include *Streptococcus thermophilus* and different species of *Lactobacillus* as *Lb. delbrueckii* and *Lb. helveticus* [10].

Cheddar cheese, soft cheeses, and most Dutch cheeses are produced with mesophilic starters. Starter culture have to meet different expectations as lactic acid production, reduction of pH, to be resistant to bacteriophages, inhibition of pathogenic and spoilage bacteria, contribute to the flavour, texture formation and have health benefit. Depending on the type of cheese the starter culture contributes to the development of nature of the cheese due to lipolysis, proteolysis and autolysis of the cells [10, 11].

Beside the lactic acid bacteria *Debaryomyces hansenii* and *Yarrowia lipolytica* yeasts are promising cheese starter cultures affecting the cheese maturation [12]. An example of molds as secondary starter culture are *Penicillium roqueforti* and *Penicillium camemberti* used in the production of blue- veined cheese. They tolerate the high and medium salinity conditions and possess high proteolytic activity [13]. There are traditional processed cheeses without starter culture, for example Spanish varieties. In these case at the beginning the high proportion of microbiota came from the milks' microbiota as lactococci.

4. The characteristic cheese microbiota

Dairy products are perishable foods providing a good medium for microbial growth. Foodborne pathogens as *Staphylococcus aureus* in raw milk can be originated from infected lactating animals, undesirable milking conditions that lead to milk contamination with animal faeces and other materials. Incorrectly handled equipment, cleaning, sanitizing equipment, water or air bleeds can provoke a multitude of contamination with various microorganisms [14].

Superior varieties of cheeses as Emmental, Gruyère, Comté, Parmigiano Reggiano, Reblochon and Roquefort are made from raw milk. The microbial safety of these products is ensured by the heat treatment at a higher temperature > 50°C, with duration of 1 h. The challenge of this type of cheeses represents the pathogen bacteria from raw milk that can grow during the process [15].

The different types of cheeses are characterized by diverse microbiota. The raw milk cheeses (soft and hard, short or long maturated, starter added or nonstarter added cheeses, and ripened in brine or unbrined cheeses) own their distinct microflora.

Gram-negative bacteria and salt-tolerant bacteria belonging to different genus, including *Halomonas, Vibrio*, and *Hafnia* spp., and salinity tolerant bacteria, including *Arthobacter, Micrococcus, Brevibacterium* and *Brachybacterium* spp., have been characteristic bacteria of raw milk cheese surface microbiota [16].

The environmental conditions of the dairy farm have an impact on the microflora of the raw milk and therefore influence manufacturing. The main spore formers in milk are species of *Clostridium*, *Bacillus* and *Geobacillus* genus. These bacteria are the follows: *C. sporogenes*, *C. butyricum*, *C. tyrobutyricum*, *B. cereus*, *B. sporothermodurans*, *G. stearothermophilus.* For the occurrence of defects in cheese as off-flavors, late blowing is responsible *C. tyrobutyricum.*

Bacteria species of *Enterococcus* are widespread in dairy products affecting negatively or positively the flavor development as the result of hydrolytic processes, proteolysis and peptidolysis. Their origin is related to raw material or to starter culture composition. In southern European artisanal cheeses, *Enterococci*'s number is higher than 10^7 CFU g^{-1}. This bacteria is an opportunistic nosocomial pathogen. Also, *Enterococci* could play a role in antibiotic resistance transfer and hospital-acquired infections. Consequently, the occurrence of this in dairy products is suspicious [2, 17]. In Gouda cheese the *Ec. malodoratus*, as its name suggest lead to the development of bad flavours [15]. *Streptococcus* species also can be isolated from milk [17].

Fox et al. [15] reported that in cheese the non-lactic acid bacteria are bacteria belonging to genera *Agrococcus*, *Arthrobacter, Brevibacterium, Brachybacterium, Corynebacterium, Microbacterium, Propionibacterium, Pediococcus, Micrococcus, Kocuria, Kytococcus, Staphylococcus.*

Aged cheese microbiota includes non-starter lactic acid bacteria. The predominantly found bacteria are facultatively heterofermentative as *Lactobacillus casei* and *L. paracasei*. It occurs occasionally *Pediococcus* spp. and obligately heterofermentative *Lactobacillus* spp., such as, *L. brevis* and *L. fermentum*.

Non-starter lactic acid bacteria origin is the raw material and/or the processing environment. A little part of the lactobacilli are resistant to the thermal treatment, pasteurization, for example in Emmental cheese obtained from raw milk can survive the 52°C temperature.

Majority of non-starter lactic acid bacteria can be characterized as: facultative anaerobes, salt-and acid-tolerant and able to grow in cheese. The involvement of these bacteria in the production of cheese flavour is unclear [2].

The number of *Micrococcus* and *Staphylococcus* is higher than 10^6 CFU g^{-1} in different types of cheese. It was found on the surface of soft smear-ripened, semi hard and hard cheeses. *Staphylococcus saprophyticus*, *S. equorum*, *S. vitulus*, and *S. xylosus* were described as main staphylococci occurred in cheese. *Kocuria*, *Nesterenkonia*, *Kytococcus*, *Dermacoccus*, and *Micrococcus* were found in cheese. *Kocuria rosea*, *K. varians*, *Dermacoccus sedantarius*, *Micrococcus lylae*, and *M. luteus isolated from* the surface of cheese. The function of *Micrococcus* and *Staphylococcus* in cheese is unclear, but many of them are related to the production of proteinases and lipases [2].

Coryneform bacteria, especially *Brevibacterium linens* is involved in the red or orange colour formation of the surface in smear-ripened cheeses. Therefore, intentionally is applied on the surface of cheeses after brining. It was proved that *Arthrobacter*, *Agrococcus*, *Brachybacterium*, *Corynebacterium*, and *Microbacterium* spp. have an important role. Origin of these corynebacteria is the brine and shelving. It is believed that staphylococci, coryneforms and micrococci, can be originated from skin due to manual handling, comprising the microflora of the skin [2].

Swiss-type cheese as Emmental and Comte´ contain propionic acid bacteria, *Propionibacterium freudenreichii*. Their pivotal role is the metabolization of lactate produced by the starter bacteria to propionate, acetate, and CO_2 [2].

Cheese can be a reservoir and carrier of probiotic bacteria for human. Due to the low redox potential, this product is a protective environment for probiotics. Their resistance to changes in pH represents an advantage in the acidic environment of gastrointestinal track. Probiotic bacteria applied in cheese production include *Lactobacillus acidophilus*, *L. casei*, *L. delbrueckii* subsp. *bulgaricus, L. johnsonii*, *L. reuteri, L. rhamnosus, Bifidobacterum bifidum, B. animalis, B. brevis, B. longum, B. infantis* [7].

Escherichia coli and coliforms are hygienic indicator bacteria. There are limit values for cheeses as well as for raw-milk cheeses.

The level of indicator microorganisms is less than 10 CFU/g and not more than 100 CFU/g, these representing the limit values in many countries.

In those instances, where cheeses had high levels of indicator bacteria, most studies attributed these to the use of poor-quality raw milk which contained high levels of indicator flora or to unsanitary conditions, or both.

The occurrence of coliforms in dairy products reveal low grade raw milk, poor hygienic conditions under production process. Coliforms could be avoided by respecting personal hygiene and sanitation process.

Human pathogens including *Salmonella* spp., *Streptococcus* spp., *L. monocytogenes*, *Helicobacter pylori*, *Campylobacter* spp., *Escherichia coli*, *Coxiella burnetti*, *Mycobacterium* spp., *Brucella* spp., tick-borne encephalitis virus (TBEV), *Staphylococcus aureus*, *Arcanobacter pyogenes*, *Toxoplasma gondii*, *Bacillus cereus*, *Leptospira*, *Clostridium* spp., and *Yersinia enterocolitica* can be present in raw milk cheeses. The inadequate fermentation conditions contribute to the growth and development of these pathogens [16].

At the origin of unwanted yeasts and molds appearance in cheeses could stand environmental factors, improper processing conditions or infested starter culture. Several yeast could affect negatively (sometimes positively) the cheese. These microbes can be involved in taste and flavour development, affecting the quality attributes also, causing spoilage and release unpleasant aroma compounds. Clinically relevant yeast was detected in different type of cheese as *Candida albicans*, *C. glabrata*, *C. krusei* and *C. tropicalis* [3].

The advanced hygiene and quality control contributed to the decrease of gas production by yeasts during cheese manufacturing [2].

A diverse group of yeasts represent a high proportion of surface microflora of different type of smear- and mold-ripened cheeses, including Comte´, Tilsit, Limburger, Blue, and Camembert. These microorganisms level can reach 10^6–10^7 CFU/g and are characterized by tolerance to low pH and high salt concentrations.

The most prevalent yeast species isolated from cheese were *Galactomyces candidus, C. parapsilosis, C. sake, C. batistae, Debaryomyces hansenii, Geotrichum candidum, Kluyveromyces lactis, K. marxianus, Saccharomyces cerevisiae, Yarrowia lipolytica, Pichia kudriavzevii* [2, 3, 7].

The most common molds in cheese include *Penicillium roqueforti* in Blue cheeses (e.g., Stilton, Roquefort, and Gorganzola) and *Penicillium camemberti* in surface mold-ripened cheeses (e.g., Camembert and Brie) [2, 3, 7]. Molds isolated from various retail cheese were: *Penicillium verrucosum*, *P. paneum*, *P. brevicompactum*, *P. gladioli*, *Aspergillus niger*, *A. flavus*, *A. phoenicis*, *Cladosporium sphaerospermum*, *C. cladosporioides*, *Eurotium rubrum* [3]. The presence of *P. commune*, which is closely related to *P. camemberti*, can cause discoloration of cheese surfaces and the development of off-flavours [15].

5. Microbiota and characteristics of traditionally processed cheeses

Artisanal cheeses represent a particular category of cheeses that is appreciated worldwide. Particular properties of products as the individual sensory aspects are related to local environment, cultural and historical features [8].

One of the challenges of artisanal cheese producers is to keep the microbial diversity of the product and exploit their beneficial effects. The traditional process, including the wooden equipment, microorganisms are the main factor in food safety issues [18].

In raw milk, over 400 species of lactic acid bacteria, Gram-positive and negative, catalase-positive bacteria, yeast and mould have been found. In cheese cores the diversity is not very large, typical lactic acid bacteria are the predominant microorganisms, whereas the surface of cheeses is rich in several species of bacteria, moulds and yeast.

Diversity within the cheeses results from considerable variation in distribution and growth of the same bacteria in the various cheese. Raw milk cheeses compared to the others are characterized by an intense and rich flavour. This results from the native microflora of the raw material and is not a character of cheeses made from treated milk [18].

The artisanal cheeses possess a heterogeneous microbiota characterized by a typical population dynamic. The unique microflora is associated with geographical indication for that region [19].

Perhaps natural milk cultures are used for italian traditional cheeses including Asiago, Pecorino Siciliano, Canestrato Pugliese, Castelmagno and Fossa. The origin of these cultures is the raw milk. The processing comprises a heat treatment of the raw milk for 15–30 min on 60–65°C, fast cooling and back slopping technique. Thermophilic species of that type of cultures are *Streptococcus thermophilus, Streptococcus macedonicus*, *Enterococcus*, and *Lactobacillus* spp. The whey based cultures contain aciduric and thermophilic strains. Among the cultivable bacteria *Lactobacillus helveticus, Lactobacillus delbrueckii* subsp. *lactis*, *Lactobacillus fermentum*, and *Str. thermophilus* were identified [20].

Central and Southern Italian cheese is “Pecorino” with PDO status, processed traditionally. The raw milk and natural microflora contribute to the development of typical organoleptic attributes. The main representants are *Enterococcus faecium*, *E. faecalis*, *Lactobacillus plantarum*, *L. brevis*, *Lactococcus lactis* subsp. *cremoris* [19].

In traditional cheese production crucial role have the autochthonous starter cultures, the main steps of manufacturing are more affected by it, but the final product are less. This refers only to that cheeses that are processed in the same geographical area and with settled circumstances. Numerous parameters have effect on the specific characteristics of cheese produced in different locations. This is case of PDO Pecorino Siciliano that is manufactured in different places of Sicily. There are some common parameters that must be applied during production as raw ewe’s milk, lamb rennet paste, wooden equipment and no added starter.

The inconstancy in the raw milk quality and the variable autochthonous microbiota determine the alteration of the attributes as organoleptic characteristics of artisanal cheese manufactured in different locations. With the application of lactic acid bacteria, the characteristics of PDO Pecorino Siciliano cheese production could be stabilized [21].

Brazilian traditional processed cheeses differ from the commercial product in their unique characteristics. These are produced with tradition, the essence of production is transferred from one generation to the next. These type of cheese are characterized by unique organoleptic properties, typical aroma, flavour, and colour due to the typical endogenous microbiota. This microbiota includes lactic acid bacteria, yeasts, fungi that source is the raw material and the production environment. Among the lactic acid bacteria isolated from artisanal cheese are mainly *Lactobacillus plantarum* and other strains as *Lactobacillus brevis*, *Lactobacillus paracasei*, *Lactococcus lactis*, *Lactobacillus rhamnosus*, *Pediococcus pentosaceus*, *Lactobacillus curvatus*, *Lactobacillus paraplantarum*, *Lactococcus garvieae*, and *Pediococcus acidilactici* [22].

The traditionally processed cheeses from raw milk are very popular and preferred by the local population, therefor is a great need for the microbiological safety of this product. For this reason, there are characterized lactic acid bacteria with multi effect traits as anti-listerial effect isolated from artisanal cheeses. *Enterococcus faecalis* (1–37, 2–49, 2–388 and 1–400, *Lactobacillus brevi*s 2–392, *Lactobacillus plantarum* 1–399 a4) have antibacterial effect against *L. monocytogenes* growth during the low temperature storage soft cheese and decrease the ripening time of semi-hard cheeses aged at ambient temperature [23].

Based on Margalho et al. [24] research from the 220 lactic acid bacteria originated from artisanal Brasil cheese twenty-two isolates had probiotic traits and role in biopreservation. These strains are characterized by resistance to gastric acids as low pH values, bile salts, and also possess good adhesion properties, mostly to enterocyte-like cells (Caco-2). One of the lactic acid bacteria, namely *Lactobacillus plantarum* (1QB77) was able to produce bacteriocin and exhibited inhibition effect against enterotoxigenic *S aureus* and *L. monocytogenes.* It was able to survive the processing conditions also.

Valuable source of microorganisms with specific characteristics are the artisanal cheese due to their high biodiversity influenced by the local ecosystem.

The Russian artisanal Buryatia cottage cheese is characterized by typical taste and flavour determined by the numerous lactic acid bacteria. From the 62 isolated and assayed lactic acid bacteria of Russian cheese belonged to different 6 genera, 14 species/subspecies. The full-length 16S rRNA sequencing revealed a high diversity of bacteria including: *Lc. lactis*, *S. thermophilus*, *Lactococcus raffinolactis*, *Acetobacter cibinongensis*, *Lactobacillus helveticus*, *Klebsiella pneumonia*, *Acinetobacter johnsonii* and *Klebsiella oxytoca* [25].

Northern Moroccan artisanal cheese is characteristic of rural regions. Due to the raw material, the goat's milk, this cheese possesses unique organoleptic properties. Galiou et al. [26] revealed that the microbiota of 28 Northern Moroccan artisanal cheeses included *Enterococcus* spp. (249 isolates), *Lactococcus lactis* (36), *Lactobacillus plantarum* (7) and *L. paracasei* (2). Besides, there were still present coliforms, yeasts, microscopic fungi and in some sample *Listeria monocytogenes.*

The Spanish Manchego is traditionally processed product, obtained with enzymatic coagulation. It is a maturated, uncooked and pressed cheese, characterized by high fat content. The 248 strains of predominant lactic acid bacteria obtained from manufacture and ripening process of two dairies included species as *L. plantarum*, *L. brevis*, *L. paracasei* subsp. *paracasei, L. fermentum, L. pentosus, L. acidophilus* and *L. curvatus*. In samples from both dairies, the species *L. plantarum, L. brevis* and *L. paracasei* subsp. *paracasei* dominated during ripening [27].

Portuguese Serpa cheese is designated as Protected Designation of Origin (PDO) cheese that is characterized also, by particular appreciated aroma and flavour. The organoleptic attributes have resulted from the not pasteurized raw ovine milk components and the addition of extracts of *Cynara cardunculus* L.. Based on literature data, the most relevant bacteria species of Serpa cheese are *Lactobacillus* spp. (*L. paracasei, L. plantarum*, *L. brevis, L. pentosus, L. curvatus*), *Leuconostoc mesenteroides, Lactococcus* lactis, coliforms, *Enterococcus faecalis*, *E. faecium*, *E. hirae*, *Hafnia alvei*, *Galactomyces* spp., *Debaryomyces hansenii*, *Kluyveromyces lactis*, *Candida zeylanoides*, *Pichia fermentans*, *Cryptococcus oeirensis*, *Yarrowia lypolytica* [28].

Northen French PDO Maroilles cheese is a soft type cheese with washed rind. The duration of maturation are 4–6 weeks according to the size, while the rinds are washed weakly with salt water. It is characterized by a surface microbiota *Geotricum candidum*, *Kluyveryomyces lactis, Brevibacterium linens* and *B. aurantiacum.* The industrially and traditionally processed cheeses differ in microorganism's composition, total mesophilic flora, rate of lactic acid bacteria and *Enterobacteriaceae*. The occurrence of *Enterobacteriaceae* in traditionally processed Maroilles reveals unhygienic conditions during manufacturing process [1].

North-western Iranian Koopeh is obtained from raw sheep or cow milk in rural households, and the ripening is under the soil for four months. This cheese is semi-hard type, characterized by high fat content, by buttery and pungent flavour. It has beneficial effects as antioxidant and antimicrobial activities attributed to antioxidant peptides and aromatic herbs content (*Allium ampeloprasum, Cuminum cyminum, Nigella sativa, Thymes vulgaris*) [29]. The non-starter lactic acid bacteria are the dominant microorganisms of the ripening steps.

The artisanal Mexican cheese, named Chiapas cream cheese is semi-ripened type. Organoleptic characteristics include the creamy and acidic taste. The predominant acidifying bacteria are the endogenous lactic acid bacteria adapted to the local climatic, tropical conditions. The bacterial strains originated from this type of cheese are characterized by high antioxidant capacity and proteolytic activity. It was showed that these bacteria could to produce bioactive compounds with role in functional foods, resulted from hydrolysis of proteins and also contain angiotensin-converting enzyme [30].

The PDO Parmigiano Reggiano traditional cheese is obtained from raw cow's milk and added whey starters obtained with back sloping. It is a long ripened type of cheese, produced in defined geographical areas (Po river's valley). The biochemical characteristics of cheese matrix are the result of the adventitious bacterial dynamics.

In cheese post brining process, that can take a 24 month different lactic acid bacteria were detected applying cultural and non-cultural methods: *Lacticaseibacillus casei, Lacticaseibacillus rhamnosus, Limosilactobacillus fermentum, Pediococcus acidilactici, Levilactobacillus brevis, Lacticaseibacillus paracasei, Lactobacillus helveticus, Lactobacillus delbrueckii ssp. lactis, Lentilactobacillus buchneri, Schleiferilactobacillus harbiniensis, Lentilactobacillus kefiri, Lactococcus lactis, Streptococcus suis, S. thermophilus* and *Lentilactobacillus parabuchneri* [31].

Merchán et al. [32] reported that the 149 yeast from Extremadura region traditional cheese, identified by ISSR-PCR technique included *D. hansenii*, *D. vindobonensis*, *K. marxianus*, *K. lactis*, *C. cabralensis*, *C. pararugosa*, *C. zeylanoides*, *Y. lipolytica*, *P. fermentans*, *P. cactophila*, *P. kudriazevii* and *P. jadinii*. *K. lactis* and *P. fermentans* were characterized by tolerance to the most restrictive stomach and bile salts stress conditions. *P. fermentans* was characterized by antimicrobial, antioxidant, hydrophobicity and auto-aggregation ability. *Kluyveromyces* spp. strains posed antimicrobial activity. The studied traditional cheese also contained strains with probiotic features as different strains of *P. fermentans*, *P. kudriavzevii*, K. *marxianus, D. hansenii, Y. lipolytica*.

The turkish artisanal cheese Tulum is a semi-hard cheese, ripened in goat skin and characterized by a unique flavour and microbial diversity. The texture is porous thus permeable to air and water. The microorganisms identified by 16 S rRNA sequencing from this cheese during ripening included: *Enterococcus faecium, E. hirae, E. durans, E. faecalis, Lactobacillus plantarum, Lactobacillus paracasei, Lactobacillus helveticus, Lactobacillus brevis, Lactococcus lactis, Streptococcus gallolyticus, Weisella paramesenteroides, Eubacterium rangiferina*. During ripening the level of Lactococci and Streptococci decreased. At the end of the maturation the DGGE profile of the bacterial community formed from *Streptococcus gallolyticus*, *Lactobacillus delbrueckii* subsp. *bulgaricus*, *Streptococcus lutetiensis*, *Streptococcus* spp., and *Enterococcus hirae*. At the end of the maturation, the most prevalent species were *Enterococcus* spp. and *Streptococcus* spp. [33].

The MALDI-TOF MS method resulted that greek artisanal cheese non-starter lactic acid bacterial community consisted of *Lactobacillus brevis, Lactobacillus plantarum, Lactococcus lactis*, *Leuconostoc mesenteroides, Lactobacillus paracasei*, *Lactobacillus rhamnosus*, *Pediococcus pentosaceus* and *Enterococcus faecium.* The first two were the most predominant species. This type of cheese is obtained from raw sheep and goat milk without added starter cultures [34].

The PDO Robiola di Roccaverano soft Italian cheese is processed from raw goat milk with natural starters. From fresh and matured cheese samples the following microorganisms were identified *Lactococcus lactis* subsp. *cremoris, Lactococcus lactis* subsp. *lactis, Leuconostoc mesenteroides, Lactobacillus plantarum, Leuconostoc pseudomesenteroides, Lactobacillus brevis, Leuconostoc citreum)* and microscopic fungi (*Geotrichum candidum, Kluyveromyces marxianus, Saturnispora silvae, Yarrowia lipolytica, Kluyveromyces lactis, Trichosporon coremiiforme* [35].

Kashkaval is an artisanal pasta filata cheese, obtained from sheep's raw milk without added starters, but now days appeared the industrialized form of this cheese using pasteurised milk with an added commercial starter. Microbiology studies of this raw milk cheese revealed that the dominant bacteria were the mesophilic non-starter lactobacilli and enterococci. In the pasteurized milk cheese *Lactococcus lactis* starter were dominant [36].

Nicastrese goat's cheese is a Southern Italian traditional product obtain from raw goat milk coagulated with lamb rennet, without any added starter cultures. The local aromatic Mediterranean plants impart unique sensorial characteristics to the milk. The typical flavour of this type of cheese is resulted from the macronutrient transformations during the maturation time. The autochthonous microbiota is responsible for the biochemistry of ripening leading to unique sensorial features. The following bacteria were detected from Nicastrese goat's cheese samples: *Lactococcus lactis* subsp. *lactis*, *Lactobacillus plantarum*, *Enterococcus faecalis*, *Lactococcus lactis* subsp. *cremoris*, *Lactobacillus casei*, *Leuconostoc mesenteroides*, *Lactobacillus brevis*, *Enterococcus faecium*, *Lactococcus garvieae*, *Enterococcus durans*, *Enterococcus hirae* and *Streptococcus thermophilus* [37].

The southern Mexican Poro de Tabasco cheese is an artisanal product obtained from raw whole cow's milk and with natural whey starters. The unique character is attributed to salting, pressing process and the use of a paraffin coating. It is a hard fresh type of cheese with full-fat content. Dominant bacteria strains of this type of cheese include *Lactobacillus* spp., *Lactococcus* spp., and *Streptococcus* spp.. The microbial level it was reduced during the processing. The final products were free of *E. coli*, *Salmonella* spp., and *L. monocytogenes*. Meanwhile, *S. aureus* and its toxins were detected in few samples. This could be eliminated with heat treatment and good manufacturing practices [38].

11 artisanal Irish cheeses (soft, hard, and semi hard) microbiome analysis by high throughput DNA sequencing revealed the presence of *Lactococcus*, *Leuconostoc*, and *Lactobacillus. Corynebacterium*, *Facklamia*, *Flavobacterium*, and *Cronobacter* were found in rind cheeses. The lactococci proportion were higher in naturally ripened rinds compared to smear and washed rinds [39].

A traditional fried cottage cheese contained strains of *Galactomyces geotrichum* molds responsible for the production of polyunsaturated fatty acids. This microscopic fungi is part of natural microflora and might have a role in omega-3 lipid production in foods and cheese [40].

The predominant bacteria of Fresh ovine cheese were represented by *Lactococcus lactis* strains added as starter culture. The characteristic of this type of cheese is given by the saffron (*Crocus sativus* L.) applied as additive. The total and lactococci bacteria level were high after storage of the product. Due to saffron content the cheese possesses antibacterial activity toward coliforms and enterococci [41].

The Calabrian (Italy) artisanal cheese lactic acid bacteria possess antibacterial effect. This group of bacteria contributed to the growth reduction of *L. monocytogenes* in vitro conditions [42].

The component of the microbiota of cheeses is influenced by the manufacture conditions as the used cheese making vessels. Generally, in artisanal cheese production, the wood is a representing material, that is a natural reservoir for microorganisms. It was showed that in these circumstances the lactic acid bacteria level is high.

In the maturation process the shelves made from wood also are responsible for appearance of coryneform bacteria, moulds, and yeasts [17].

Gram-negative bacteria strains belonging to different families were detected on the surface of maturated soft cheese. These were: *Enterobacteriaceae, Moraxellaceae, Pseudoalteromonadaceae, Pseudomonadaceae, Sphingobacteriaceae* and *Vibrionaceae* spp. The Gram-negative bacteria were attributed to undesirable and defective texture and flavour. This defects resulted from extracellular proteolytic and lipolytic enzymes activities of bacteria. An example is the occurrence of *Pseudomonas* spp. that influence negatively the organoleptic characteristics of cheese due to release of ethyl esters and alcohols. It is supposed that the Gram-negative bacteria belonging to natural microbiome affecting positively the product technological and sensorial features. A strain of *Proteus* spp. closely related to *P. hauseri* and strains of *Psychrobacter* spp. have been contributed to flavour development of smear type cheeses. In French traditional cheeses were detected Gram-negative bacterial strains with double effect. There were bacteria strains characterised by no tolerance to acidic conditions in simulated gastro-intestinal stress, to human serum and showed no adherence to Caco-2 cell and were no virulence. In contrast to *Hafnia paralvei* 920, *Proteus* spp. (close to *P. hauseri*) UCMA 3780, *Providencia heimbachae* GR4, and *Morganella morganii* 3A2A that were toxic in *in vitro* conditions possessing virulence factors [43].

6. Occurrence of allochthonous microorganisms in traditionally processed cheese

The detection of allochthonous bacteria and microscopic fungi in raw milk cheese raise food safety issues because those products can harbour diverse spoilage and pathogenic microbes. We evaluated the microbial quality of 13 traditionally processed cheese from northeastern region of Transylvania. The processed cheese are made from unpasteurized cow and sheep milk. It is considered that the milk quality is affected by grazed grass, the special feeding strategy in that region.

The detection of the various pathogen bacteria was carried out by cultivation methods on different selective media: for the detection and enumeration of *Escherichia coli* TBX Chromo Agar was used, ChromoBio®Coliform agar was applied for the detection of coliforms, for the detection of *Salmonella* spp. Brilliance ™ Salmonella Agar Base was used, for the detection of *Shigella* spp. Salmonella Shigella (SS) Modified Agar was applied, for the detection of *Campylobacter jejuni* Campylobacter Blood-Free Selective Agar Base Modified CCDA was used, for *Yersinia enterocolitica* detection the Yersinia Selective Agar Base was used, for *Listeria monocytogenes* detection Listeria mono Differential Agar Base was used, for *Bacillus cereus* detection ChromoBio®Cereus Base was used, for *C. perfringens* Clostridial Differential Broth was used, for the detection of *Pseudomonas aeruginosa* Pseudomonas Isolation Agar Base was applied. The detection of microscopic fungi was carried out by spread plate method on ChromoBio®Candida and Czapek Dox Agar [44, 45].

Isolation of the most representative bacteria with high count and characteristic colony morphology was followed by identification of strains on species-level by 16 rDNA sequencing method. Genomic DNA isolation of isolated bacterial strains from the selective mediums was realized with AccuPrep® Genomic DNA Extraction Kit from Bioneer, according to the manufacturer's protocol. A part of the bacterial 16S rDNA gene was amplified with universal primers 27f and 1492r (5' AGAGTTTGATCMTGGCTCAG 3', 5' TACGGYTACCTTGTTACGACTT 3').

The PCR products were purified using *GeneJET PCR Purification Kit* (*Thermo Scientific*, K070) and sequencing was carried out [46].

Our results of the microbiological quality of the assayed artisanal cheese showed diverse values. The highest total count of coliforms, *Escherichia coli* and *Salmonella* spp. was detected in the cheese 7 sample (**Table 1**). The mean number of *Campylobacter* jejuni was 10^3 CFU/g and 10^2–10^3 CFU/g in the case of *Yersinia enterocolitica.* The highest level was detected in the case of cheese 10 sample $8{\cdot}10^4$ CFU/g *C. jejuni* and $4.1{\cdot}10^4$ CFU/g *Y. enterocolitica.* The highest number of *Listeria monocytogenes* was detected in cheese 7 sample $7.5{\cdot}10^5$ CFU/g, in case of *Staphylococcus aureus* the highest count was $1.9{\cdot}10^7$ CFU/g in sample cheese 6. The highest level of *Bacillus cereus* was $2{\cdot}10^5$ CFU/g in sample cheese 7 (**Table 1**). The *Clostridium perfringens* was detected in two samples cheese 3, cheese 8 samples.

The count of microscopic filamentous fungi in the assayed samples obtained by us varied 10^2–10^3 CFU/g. The highest load was found in cheese 4 sample $5{\cdot}10^4$ CFU/g. The maximum value detected in the case of yeasts was $4.5 \cdot 10^5$ CFU/g in cheese 9 sample (**Table 2**). The developed filamentous fungi belonged to *Aspergillus* and *Penicillium* genera (**Figure 1**).

The identified 27 bacterial strains isolated on various selective media according to 16S rDNA sequence analysis include: *Stenotrophomonas maltophilia*, *Escherichia coli*, *Staphylococcus equorum* subsp. *equorum*, *Staphylococcus equorum* subsp. *linens*, *Klebsiella michiganensis*, *Halomonas alkaliphila*, *Proteus vulgaris*, *Shigella flexneri*, *Enterococcus durans, Enterococcus faecalis, Bacillus licheniformis, Kocuria rhizophila, Hafnia paralvei* (**Figure 2**).

The occurrence of *Stenotrophomonas maltophilia* in cheese samples originated from different regions of France was reported by Coton et al. 2012 [47]. This multidrug-resistant bacteria causing nosocomial infections, can be found in different environments as aquatic or humid, in natural soil or plant rhizosphere [48, 49]. In immunocompromised patients this bacteria was associated with pneumonia and bacteremia. Also can cause fatal hemorrhagic pneumonia. It has been found that the StmPr1 gene is involved in pathogenicity [48]. Biofilm formation (the presence of genes rpfF and smf-1 genes) was considered that play role in virulence [50].

In dairy products the presence of *Escherichia coli* indicates fecal contamination. It was detected in Brazilian raw milk that could be appear in Minas Frescal-type cheeses produced from raw milk. The risk could be eliminated with the heat treatment of milk [51]. Metz et al. 2020 [52] reported that *E. coli* and coliforms (detected in raw milk) number, rise during the early steps of processing. At the end of the cheese making, including fermentation and ripening, their level decline by numerous logs or the final product is free of this bacteria. Contaminated raw milk of domestic animals and dairy products could be the reservoir of verotoxin-producing *Escherichia coli* (VTEC). This bacteria can cause severe infections also. Outbreaks of *Escherichia coli* O157:H7 was related to contaminated raw milk. It was reported that *E. coli* O157 was growth in initial phase of preparation of Fontina PDO cheese. The heat treatment of curd cooking not affected the bacteria. Only during ripening decreased the level of *E. coli.* It was considered that the a_w reduction and the lactic acid could contribute to the reduction of *E. coli* O157 level [53]. In Cacioricotta goat cheese *E. coli* O157:H7 could survive after 90 days of ripening. This fact gives reason for that the low level contaminated raw milk represents a threat to consumers [54].

The plasmid-mediated colistin resistance gene in *E. coli* from hard cheese represents a high risk factor because cheese is consumed without heat treatment [55]. During manufacturer process of raw milk cheese only the low temperature conditions of maturation affected the growth negatively of shigatoxigenic *E. coli* and *L. monocytogenes* [56].

Cheese sample	*Escherichia coli*	Coliforms	*Salmonella* spp.	*Shigella* spp.	*Campylobacter jejuni*	*Yersinia enterocolitica*	*Listeria monocytogenes*	*Staphylococcus aureus*	*Bacillus cereus*
Cheese 1	$2.7\cdot10^{2}$	$2.4\cdot10^{4}$	$1\cdot10$	<10	$7.6\cdot10^{3}$	$5.4\cdot10^{2}$	$1.9\cdot10^{4}$	$2.9\cdot10^{3}$	<10
Cheese 2	<10	$9\cdot10^{3}$	$6\cdot10$	<10	<10	$4\cdot10$	$2\cdot10^{5}$	$9.9\cdot10^{6}$	$1.5\cdot10^{5}$
Cheese 3	$9\cdot10$	$8\cdot10^{2}$	$1.3\cdot10^{2}$	<10	$2.5\cdot10^{3}$	<10	$9.1\cdot10^{3}$	$1.15\cdot10^{3}$	<10
Cheese 4	<10	<10	<10	<10	$1.3\cdot10^{3}$	$1\cdot10^{2}$	$3.4\cdot10^{4}$	$5.1\cdot10^{5}$	<10
Cheese 5	<10	<10	<10	<10	<10	<10	$1\cdot10^{3}$	$5\cdot10^{4}$	$1\cdot10^{3}$
Cheese 6	$1.2\cdot10^{3}$	<10	$4.6\cdot10^{2}$	<10	$5\cdot10^{3}$	$7.2\cdot10^{3}$	$4\cdot10^{3}$	$1.9\cdot10^{7}$	$2\cdot10^{4}$
Cheese 7	$6.8\cdot10^{3}$	$8\cdot10^{4}$	$3.2\cdot10^{4}$	<10	$2.6\cdot10^{3}$	$1\cdot10^{2}$	$7.5\cdot10^{5}$	$1.22\cdot10^{5}$	$2\cdot10^{5}$
Cheese 8	<10	<10	<10	<10	$1.8\cdot10^{3}$	$1.57\cdot10^{3}$	$7.3\cdot10^{4}$	$1.55\cdot10^{7}$	<10
Cheese 9	<10	$2.6\cdot10^{4}$	<10	<10	$7\cdot10$	$2.79\cdot10^{2}$	$8.1\cdot10^{3}$	$1\cdot10^{3}$	$4\cdot10^{5}$
Cheese 10	<10	$2\cdot10^{2}$	<10	<10	$8\cdot10^{4}$	$4.1\cdot10^{4}$	$2.51\cdot10^{3}$	$3\cdot10^{2}$	<10
Cheese 11	$1\cdot10^{2}$	$2\cdot10^{2}$	<10	<10	$1\cdot10^{4}$	<10	$5.6\cdot10^{3}$	$1\cdot10^{3}$	$9.1\cdot10^{4}$
Cheese 12	$1\cdot10^{3}$	$1.7\cdot10^{3}$	<10	<10	$3\cdot10$	<10	$9.6\cdot10^{3}$	$4.2\cdot10^{3}$	<10
Cheese 13	<10	<10	<10	<10	<10	<10	$3.8\cdot10^{2}$	$7\cdot10^{2}$	$2\cdot10^{3}$

Table 1.
Microbial load of cheese samples (CFU/g).

Cheese sample	Molds	*Candida* spp.
Cheese 1	$4.3{\cdot}10^2$	<10
Cheese 2	$7{\cdot}10^2$	<10
Cheese 3	$3.5{\cdot}10^3$	<10
Cheese 4	$5{\cdot}10^4$	<10
Cheese 5	$9{\cdot}10^3$	<10
Cheese 6	$2.4{\cdot}10^4$	$3.5{\cdot}10^3$
Cheese 7	$1.2{\cdot}10^2$	$1.3{\cdot}10^2$
Cheese 8	$9{\cdot}10^2$	$7.3{\cdot}10^2$
Cheese 9	$7{\cdot}10^2$	$4.5{\cdot}10^5$
Cheese 10	$7{\cdot}10^2$	<10
Cheese 11	$5{\cdot}10$	$3.19{\cdot}10^5$
Cheese 12	0	$3.75{\cdot}10^3$
Cheese 13	0	$2.89{\cdot}10^3$

Table 2.
Occurrence of microscopic fungi in the analyzed cheese samples (CFU/g).

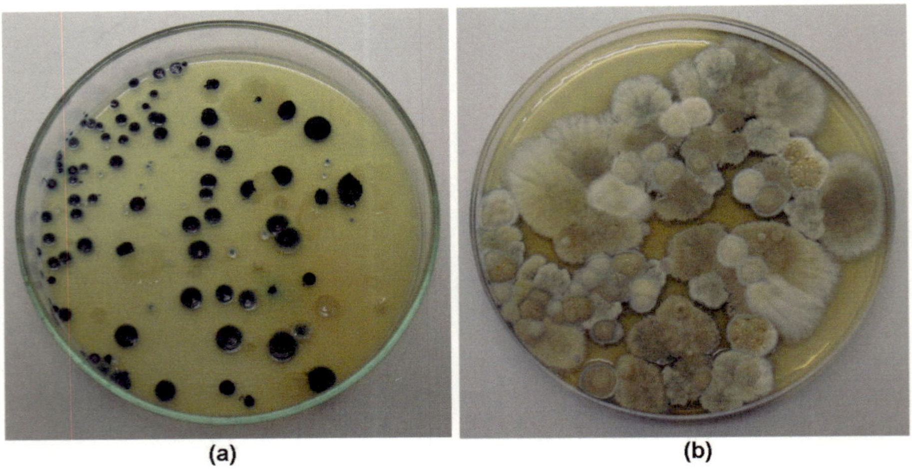

Figure 1.
Colonies of aerobic spore-forming bacteria from cheese 6 sample (A) and microscopic fungi from cheese 10 sample (B) on selective agar plates.

One of the predominant bacteria from staphylococci is *S. equorum* in cheeses owing antilisterial activity. *S. equorum* and *S. equorum* subsp. *linens* was mentioned as potential new starter culture [57]. *S. equorum* also characterized by high salinity tolerance (up to 25%) was the predominant species of the long maturation processes.

In traditional fermented foods staphylococci were the dominant species. These bacteria are involved in the development of sensorial features. *Staphylococcus carnosus*, *S. xylosus* and *S. equorum* with the production of low molecular-weight compounds, including aldehydes, amines, amino acids esters and free fatty acids take part in flavor development. This is possible due to various biochemical transformations as reduction of nitrates to nitrite and then to nitrous oxide, proteolysis and lipolysis [58]. With DNA–DNA hybridization method it was revealed that

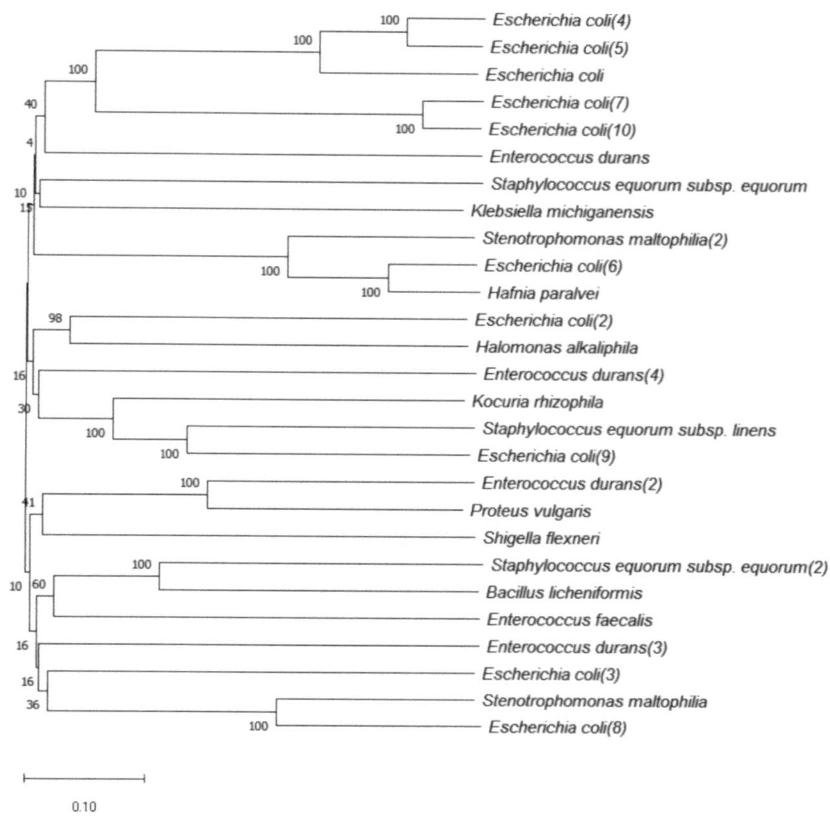

Figure 2.
Phylogenetic tree based on similarities of 16S rDNA partial sequences of bacterial isolates using the neighbor-joining method.

Staphylococcus equorum subsp. *linens* subsp. nov. and *Staphylococcus equorum* subsp. *linens* was part of the dominant microflora of surface ripened Swiss mountain raw milk cheese. Characteristics of these two strains include the ability to grow between 6 and 40°C and in salinity conditions 13% NaCl [59].

Based on literature data *Klebsiella michiganensis* originated from traditionally processed cheese is considered an emerging pathogen, causing nosocomial and bloodstream infections. It was isolated from hospital samples. Different virulence factors were identified in this bacteria as capsule locus type, plasmid replicon types, cell adherence and iron permitting acquisition. *K. michiganensis* is a multidrug resistant bacteria harboring (KPC)-3 carbapenemase gene, carbapenemase gene *bla*OXA-181 and other 11 genes encoding different antibiotic resistance. The carbapenemase gene encoding plasmid is originated from *Klebsiella pneumoniae* [60, 61]. Mitra et al. [62] reported that K. *michiganensis* strain S8 possess cadmium accumulating capacity contributing to enhance the productivity of rice.

The other identified bacteria from cheese was *Halomonas alkaliphila*. According to Sun et al. [63] is a halophilic proteobacteria with role in nitrogen cycle of marine environments, converting inorganic nitrogen compounds.

It was reported that *Proteus vulgaris* potentially could be applied to enhance the flavor based on the production of widest varieties and the largest quantities of volatile compounds on the surface of a smear-ripened cheese [64]. In surface-ripened cheeses proteobacteria as *Proteus* are responsible for the development of purple pigmentation of cheese rinds [65].

Shigella flexneri is known as a human pathogen infecting the colonic epithelial cells. With reduction of intercellular tension is able to spread between cells [66].

Jamet et al. [67] reported that enterococci take part of the dominant microflora of raw milk cheeses ranging from 10^2 to 10^8 CFU/g. There are contradictory aspects regarding the *Enterococcus* lactic acid bacterial strains. It was considered that these strains of bacteria are capable of causing diseases. Studies are showing that *Enterococcus* bacterial strains are promising starter cultures owing relevant technological and probiotic characteristics. *E. durans* LM01C01 with beneficial characteristics in cream cheeses lowered the *S. aureus* contamination level [68]. *Enterococcus durans* LAB18s possess probiotic potential, antimicrobial, antioxidant characteristics and selenium bioaccumulation ability. The culture supernatant of aforementioned bacteria exhibited inhibitory effect against *Listeria monocytogenes, Escherichia coli, Salmonella typhimurium, Pseudomonas aeruginosa, Aeromonas hydrophila* and *Corynebacterium fimi*. The intracellular extract of *E. durans* LAB18s showed antibacterial effect against *L. monocytogenes, E. coli, B. cereus, Staphylococcus aureus, S. enteritidis, Ps. aeruginosa, Ae. hydrophila* and *Corynebacterium fimi* [69].

Enterococcus durans SJRP29 originated from cheese is thought to be an advantageous safe strain with practical use due to the beneficial characteristics (tolerance to 3% NaCl, good adhesion and without bile salt hydrolase activity or mucin degradation) [70]. Enterococci are involved in antibiotic resistance gene transfer among non-pathogenic and pathogenic bacteria. It has been reported that the predominant antibiotic resistant enterococci in French cheeses were *E. faecalis, E. faecium* and *E. durans.* Antibiotic resistance was showed against tetracycline, minocycline, erythromycin, kanamycin and chloramphenicol [67]. *L. monocytogenes* and *E. faecalis* were true cases of horizontal gene transfer in Minas Frescal cheese [71]. In another study it was concluded that antibiotic resistance and virulence factors can be transferred with high rate during cheese and sausage fermentation to enterococci [72].

B. licheniformis due to high resistance of environmental factors and heat treatments is involved in the biofilm formation in milk processing plants, but this can be avoiding with the pH and milk residues control [73]. *B. licheniformis* is a food spoilage and poisoning bacteria causing serious diseases as enteric disease, septicaemia, peritonitis, ophthalmitis [74]. Also these spore forming bacteria are promising probiotic strains. It was reported that *Bacillus* strains could survive the harsh environment of Requeijão cremoso Brazilian cheese [75].

Kocuria as a non-pathogenic bacteria is part of skin microbiota. Several studies reported that it was associated with peritonitis and urinary tract infection. Recently it has been reported that *Kocuria rhizophila* is causing catheter-related bacteremia with endocarditis [76].

The genus *Hafnia* belonging to Enterobacteriaceae comprise bacterial strain predominant in nature, gastrointestinal tracts of mammals and also in food products. *Hafnia* sp. and *Enterococcus faecalis* was associated with cholecystitis [77].

7. Conclusions

The microflora of traditionally processed cheese comprise of diverse and heterogeneous group of microbes. These microorganisms have specific role in product formation providing the unique characteristics of the cheese. Also, it could appear microorganisms that are emerging pathogens. For these reason in traditional manufacture process have to pay attention to personal hygiene and sanitation process. The support of autochthonous bacteria in fermented products have to be a priority to contribute to the preservation of the organoleptic characteristics typical of that geographical region.

Author details

Éva György[1]* and Éva Laslo[2]

1 Faculty of Economics, Socio-Human Sciences and Engineering, Department of Food Science, Sapientia Hungarian University of Transylvania, Cluj-Napoca, Romania

2 Faculty of Economics, Socio-Human Sciences and Engineering, Department of Bioengineering, Sapientia Hungarian University of Transylvania, Cluj-Napoca, Romania

*Address all correspondence to: gyorgyeva@uni.sapientia.ro

References

[1] Nacef M, Lelièvre-Desmas M, Drider D, Flahaut C, Chollet S. Artisanal and industrial Maroilles cheeses: Are they different? Comparison using sensory, physico-chemical and microbiological Approaches. International Dairy Journal. 2019; **89**:42-52

[2] Cogan TM, Microbiology of chees, 2011. In: Encyclopedia of Dairy Sciences (Second Edition), Fuquay JW (ed.), Academic Press, pp. 625-631

[3] Banjara N, Suhr MJ, Hallen-Adams HE. Diversity of yeast and mold species from a variety of cheese types. Current Microbiology. 2015; **70**(6):792-800

[4] Pintado M, Gomes da Cruz A, Rodrigues de Sá PBZ. Cheese Microbiology, 2015. In: Dairy Microbiology and Biochemistry Recent Developments, Özer BH, Akdemir-Evrendilek G (eds.), CRC Press, Boca Raton, pp. 113-116

[5] Araújo-Rodrigues H, Tavaria FK, dos Santos MTPG, Alvarenga N, Pintado MM. A review on microbiological and technological aspects of Serpa PDO cheese: An ovine raw milk cheese. International Dairy Journal. 2020; **100**, 104561

[6] Guinee TP. Salt in cheese, 2007. In: Cheese problems solved, McSweeney PLH (ed.), CRC Press, Cambridge, pp. 80-99

[7] Hayaloglu AA. Cheese: Microbiology of cheese. 2016, Reference Module in Food Science.

[8] Schlei, K.P., Angioletti, B.L., de Carvalho, L.F., Bertoli, S.L., Reiter, M.G.R., de Souza, C.K., Influence of temperature on microbial growth during processing of kochkäse cheese made from raw and pasteurised milk, International Dairy Journal, 2020, 109, 104786

[9] Muruzović MŽ, Mladenović KG, Čomić LR. In vitro evaluation of resistance to environmental stress by planktonic and biofilm form of lactic acid bacteria isolated from traditionally made cheese from Serbia. Food Bioscience. 2018; **23**: 54-59

[10] Altieri C, Ciuffreda E, Di Maggio B, Sinigaglia M. Lactic acid bacteria as starter cultures. Starter Cultures in Food Production. 2017; 1-15

[11] Parente E, Cogan TM, Powell IB. Starter Cultures: General Aspects. Cheese. 2017; 201-226

[12] Buzzini P, Di Mauro S, Turchetti B. Yeasts as starter cultures. Starter Cultures in Food Production. 2017; 16-49

[13] Dantigny P., Bevilacqua A. Fungal starters. Starter Cultures in Food Production. 2017; 50-63

[14] Lourenco A, Fraga-Corral M, De Colli L, Moloney M, Danaher M, Jordan K. Determination of the presence of pathogens and anthelmintic drugs in raw milk and raw milk cheeses from small scale producers in Ireland. LWT - Food Science and Technology. 2020; **130**, 109347

[15] Fox PF, Guinee TP, Cogan TM, McSweeney PLH. Fundamentals of Cheese Science, 2nd Edition, Springer, New York. 2017; pp:113, 373-381

[16] O'Sullivan O, Cotter PD. Microbiota of Raw Milk and Raw Milk Cheeses, 2017. In: Cheese. Chemistry, Physics & Microbiology, vol.1., General Aspects, 4th Edition, McSweeney PLH, Fox PF, Cotter PD, Everett DW (ed.), Academic Press, London, pp. 301-310

[17] Gagnon M, Hamelin L, Fréchette A, Dufour S, Roy D. Effect of recycled manure solids as bedding on bulk tank milk and implications for cheese microbiological quality. J. Dairy Sci. 2020; **103**:128-140

[18] Montel M-C, Buchin S, Mallet A, Delbes-Paus C, Vuitton DA, Desmasures N, Berthier F. Traditional cheeses: Rich and diverse microbiota with associated benefits. International Journal of Food Microbiology. 2014; **177**: 136-154

[19] Centi V, Matteucci F, Lepidi A, Del Gallo M, Ercole C. Microbiological and biochemical aspects of inland Pecorino Abruzzese cheese. Heliyon. 2017; **3**, e00258.

[20] Gobetti M, Neviani E, Fox P. The Cheeses of Italy: Science and Technology. 2018. Springer, Cham, pp: 68-70

[21] Gaglio R, Franciosi E, Todaro A, Guarcello R, Alfeo V, Randazzo CL, Settanni L, Todaro M. Addition of selected starter/non-starter lactic acid bacterial inoculums to stabilise PDO Pecorino Siciliano cheese production. Food Research International. 2020; **136,** 109335

[22] Margalho LP, Feliciano MD, Silva CE. Abreu JS, Piran MVF, Sant'Ana AS, Brazilian artisanal cheeses are rich and diverse sources of nonstarter lactic acid bacteria regarding technological, biopreservative, and safety properties—Insights through multivariate analysis. J. Dairy Sci. 2020; **103**:7908-7926

[23] Campagnollo FB, Margalho LP, Kamimura BA, Feliciano MD, Freire L, Lopes LS, Alvarenga VO, Cadavez VAP, Gonzales-Barron U, Schaffner DW, Selection of indigenous lactic acid bacteria presenting anti-listerial activity, and their role in reducing the maturation period and assuring the safety of traditional Brazilian cheeses. Food Microbiology. 2018; **73**:288-297

[24] Margalho LP, Jorge GP, Noleto DAP, Silva CE, Abreu JS, Piran MVF, Brocchi M, Sant'Ana AS. Bio preservation and probiotic potential of a large set of lactic acid bacteria isolated from Brazilian artisanal cheeses: From screening to *in product* approach. Microbiological Research. 2021; **242**, 126622

[25] Jin H, Mo L, Pan L, Hou Q, Li C, Darima I, Yu J. Using PacBio sequencing to investigate the bacterial microbiota of traditional Buryatian cottage cheese and comparison with Italian and Kazakhstan artisanal cheeses. J. Dairy Sci. 2018; **101**:6885-6896

[26] Galiou OE, Zantar S, Bakkali M, Laglaoui A, Centeno JA, Carballo J. Chemical and microbiological characteristics of traditional home made fresh goat cheeses from Northern Morocco. Small Ruminant Research. 2015; **129:** 108-113

[27] Sánchez I, Seseňa S, Poveda JM, Cabezas L, Palop L. Genetic diversity, dynamics, and activity of *Lactobacillus* community involved in traditional processing of artisanal Manchego cheese. International Journal of Food Microbiology. 2006; **107**:265-273

[28] Araújo-Rodrigues H, Tavaria FK, dos Santos MTPG, Alvarenga N, Pintado MM. A review on microbiological and technological aspects of Serpa PDO cheese: An ovine raw milk cheese. International Dairy Journal. 2020; **100**, 104561

[29] Banihashemi SA, Nikoo M, Ghasempour Z, Ehsani A. Bioactive peptides fractions from traditional Iranian Koopeh cheese; lactic fermentation products. Biocatalysis and Agricultural Biotechnology. 2020; **29**, 101798

[30] Gonzalez-Gonzalez CR, Machado J, Correia S, McCartney AL, Elmore JS, Jauregi P. Highly proteolytic bacteria from semi-ripened Chiapas cheese elicit angiotensin-I converting enzyme inhibition and antioxidant activity. LWT - Food Science and Technology. 2019; **111**:449-456

[31] Levante A, Bertani G, Bottari B, Bernini V, Lazzi C, Gatti M, Neviani E. How new molecular approaches have contributed to shedding light on microbial dynamics in Parmigiano Reggiano cheese. Current Opinion in Food Science. 2021; **38**:131-140

[32] Merchán AV, Benito MJ, Galván AI, de Herrera SR-MS. Identification and selection of yeast with functional properties for future application in soft paste cheese. LWT - Food Science and Technology. 2020; **124,** 109173

[33] Dermici T, Akin N, Atik DS, Özkan ER, Dertli E, Akyol İ. Lactic acid bacteria diversity and dynamics during ripening of traditional Turkish goatskin Tulum cheese produced in Mut region assessed by culturing and PCR-DGGE. LWT - Food Science and Technology. 2021; **138**, 110701

[34] Gantzias C, Lappa IK, Aerts M, Georgalaki M, Manolopoulou E, Papadimitrou K, De Brandt E, Tsakalidou E, Vandamme P. MALDI-TOF MS profiling of non-starter lactic acid bacteria from artisanal cheeses of the Greek island of Naxos. International Journal of Food Microbiology. 2020; **323**, 108586

[35] Biolcati F, Andrighetto C, Bottero MT, Dalmasso A. Microbial characterization of an artisanal production of Robiola di Roccaverano cheese. Journal of Dairy Science. 2020; **103** (5): 4056-4067

[36] Pappa EC, Kondyli E, Samelis J. Microbiological and biochemical characteristics of Kashkaval cheese produced using pasteurised or raw milk. International Dairy Journal. 2019; **89**, 60-67

[37] Pino A, Liotta L, Randazzo CL, Todaro A, Mazzaglia A, De Nardo F, Chiofalo V, Caggia C. Polyphasic approach to study physico-chemical, microbiological and sensorial characteristics of artisanal Nicastrese goat's cheese. Food Microbiology. 2018; **70**: 143-154

[38] De la Rosa-Alcaraz M de IÁ, Ortiz-Estrada ÁM, Heredia-Castro PY, Hernández-Mendoza A, Reyes-Díaz R, Vallejo-Cordoba B, González-Córdova AF. Poro de Tabasco cheese: Chemical composition and microbiological quality during its artisanal manufacturing process. J. Dairy Sci. 2020; **103** (4): 3025-3037

[39] Jonnala BRY, McSweeney PLH, Sheehan JJ, Cotter PD, Sequencing of the cheese microbiome and its relevance to industry. Frontiers in Microbiology. 2018; **9**, 1020

[40] Grygier A, Myszka K, Juzwa W, Bialas W, Rudzińska M. *Galactomyces geotrichum* mold isolated from a traditional fried cottage cheese produced omega-3 fatty acids. International Journal of Food Microbiology. 2020; **319**, 108503

[41] Aktypis A, Christodoulou ED, Manolopoulou E, Georgala A, Daferera D, Polysiou M. Fresh ovine cheese supplemented with saffron (*Crocus sativus* L.): Impact on microbiological, physicochemical, antioxidant, colour and sensory characteristics during storage. Small Ruminant Research. 2018; **167**:32-38

[42] Panebianco F, Giarratana F, Caridi A, Sidari S, De Bruno A, Giuffrida A. Lactic acid bacteria isolated from traditional Italian dairy products: activity against *Listeria monocytogenes* and modelling of microbial competition

in soft cheese. LWT - Food Science and Technology. 2021; **137**, 110446

[43] Imran M, Desmasures N, Coton M, Coton E, Le Flèche-Matéos A, Irlinger F, Delbès-Paus C, Stahl V, Montel M-C, Vernoux J-P. Safety assessment of Gram-negative bacteria associated with traditional French cheeses. Food Microbiology. 2019; **79**: 1-10

[44] Dunca S, Ailiesei O, Nimițan E, Ștefan M. Microbiologie aplicată, Editura Tehnopress, Iași, pp. 246-253.

[45] Commission Regulation (EC) No 1441/2007 of 5 December 2007 amending Regulation (EC) No 2073/2005 on microbiological criteria for foodstuffs. Official Journal of the European Union 322/13 7.12.2007.

[46] Laslo É, Mara Gy, Funkenhauzer B, Dobri E, Salamon R.V., Lányi Sz, Ábrahám B. Isolation and screening of lactic acid bacteria from naturally fermented sources with high biotechnological potential. Studia UBB Chemia. 2015; **4**: 95-106.

[47] Coton, M, Delbés-Pausb C, Irlingerc F, Desmasuresd N, Le Flechee A, Stahlf V, Montelb M-C, Cotona E. Diversity and assessment of potential risk factors of Gram-negative isolates associated with French cheeses. Food Microbiology. 2012; **29**: 88-98

[48] Imoto W, Kaneko Y, Yamada K, Kuwabara G, Yamairi K, Wataru Shibata W, Oshima K, Niki M, Nakaie K, Watanabe T, Asai K, Kawaguchi T, Kakeya H. A mouse model of rapidly progressive fatal haemorrhagic pneumonia caused by *Stenotrophomonas maltophilia*. Journal of Global Antimicrobial Resistance. 2020; **23**:450-455

[49] Ma J, Feng J, Shan Y, Zhao Y, Qiao H, Xie L, Lin X, Wang C, Chuai X. Characteristic antimicrobial resistance of clinically isolated *Stenotrophomonas maltophilia* CYZ via complete genome sequence. Journal of Global Antimicrobial Resistance. 2020, **23**:186-193

[50] Azimi A, Aslanimehr M, Yaseri M, Shadkam M, Douraghi M. Distribution of smf-1, rmlA, spgM and rpfF genes among *Stenotrophomonas maltophilia* isolates in relation to biofilm-forming capacity. Journal of Global Antimicrobial Resistance. 2020; **23**:321-326

[51] Ribeiro Júnior JC, Silva FF, Lima JBA, E. H. Ossugui EH, Teider Junior PI, Campos ACLP, Navarro A, Tamanini R, Ribeiro J, Alfieri AA, Beloti V. Molecular characterization and antimicrobial resistance of pathogenic *Escherichia coli* isolated from raw milk and Minas Frescal cheeses in Brazil. J. Dairy Sci. 2019; **102**:10850-10854

[52] Metz M, Sheehan J, Feng PCH. Use of indicator bacteria for monitoring sanitary quality of raw milk cheeses – A literature review. Food Microbiology. 2020; **85**, 103283

[53] Bellio A, Bianchi DM, Vitale N, Vernetti L, Gallina S, Decastelli L. Behavior of *Escherichia coli* O157:H7 during the manufacture and ripening of Fontina Protected Designation of Origin cheese. J. Dairy Sci. 2018; **101**:4962-4970

[54] Ioanna F, Quaglia NC, Storelli MM, Castiglia D, Goffredo E, Storelli A, De Rosa M, Normanno G, Caputi Jambrenghi A, Dambrosio A. Survival of *Escherichia coli* O157:H7 during the manufacture and ripening of Cacioricotta goat cheese. Food Microbiology.2018; **70**:200-205

[55] Ombarak RA, Awasthi SP, Hatanaka N, Yamasaki S. Detection of plasmid mediated colistin resistance *mcr-1* gene in ESBL producing *Escherichia coli* isolated from raw milk hard cheese in Egypt. International Dairy Journal. 2021; **117**,104986.

[56] Alshaibani D, Machado RAM, Calder BL, Perry JJ. Survival of *Listeria monocytogenes* and shigatoxigenic *Escherichia coli* during the production and aging of farmstead-style cheese. International Dairy Journal. 2020, **110**, 104801

[57] Blaiotta G, Ercolini D, Mauriello G, Salzano G, Villani F. Rapid and Reliable Identification of *Staphylococcus equorum* by a Species-Specific PCR Assay Targeting the *sod*A Gene. System. Appl. Microbiol., 2004; **27**, 696-702

[58] Jeong D-W, Han S, Lee J-H. Safety and technological characterization of *Staphylococcus equorum* isolates from jeotgal, a Korean high-salt-fermented seafood, for starter development. International Journal of Food Microbiology. 2014; **188**:108-115

[59] Place RB, Hiestand D, Gallmann HR, Teuber M. *Staphylococcus equorum* subsp. *linens,* subsp. nov., A Starter Culture Component for Surface Ripened Semi-Hard Cheeses. System. Appl. Microbiol. 2003; **26**:30-37

[60] Seiffert SN, Wüthrich D, Gerth Y, A. Egli A, Kohler P, Nolte O. First clinical case of KPC-3–producing *Klebsiella michiganensis* in Europe. New Microbes and New Infections. 2019; **29:** 100516

[61] King TL, Schmidt S, Thakur S, Fedorka-Cray P, Keelara S, Harden L, Essack SY. Resistome of a carbapenemase-producing novel ST232 *Klebsiella michiganensis* isolate from urban hospital effluent in South Africa. Journal of Global Antimicrobial Resistance. 2021

[62] Mitra S., Pramanika K, Ghosh PK, Sorena T, Sarkar A, Dey RS, Pandeye S, Maiti TK. Characterization of Cd-resistant *Klebsiella michiganensis* MCC3089 and its potential for rice seedling growth promotion under Cd stress. Microbiological Research, 2018; **210**:12-25

[63] Sun X, Chen B, Li O, Liu N, Xia B, Zhu L, Qu K. Toxicities of polystyrene nano- and microplastics toward marine bacterium *Halomonas alkaliphila*, Science of the Total Environment, 2018; **642:**1378-1385

[64] Deetae P, Saint-Eve A, Spinnler HE, Helinck S. Critical effect of oxygen on aroma compound production by *Proteus vulgaris*. Food Chemistry. 2011; **126**:134-139

[65] Kamelamela N, Zalesne M, Morimoto J, Robbat A, Benjamin E. Wolfe, Indigo- and indirubin-producing strains of *Proteus* and *Psychrobacter* are associated with purple rind defect in a surface-ripened cheese. Food Microbiology. 2018; **76**: 543-552

[66] Duncan-Lowey JK, Wiscovitch AL, Wood TE, Goldberg MB, Russo BC. *Shigella flexneri* Disruption of Cellular Tension Promotes Intercellular Spread. Cell reports. 2020; **33**, 108409

[67] Jamet E, Akary E, Poisson M-A, Chamba J-F, Bertrand X, Serror P. Prevalence and characterization of antibiotic resistant *Enterococcus faecalis* in French cheeses. Food Microbiology, 2012; **31**:191-198

[68] Estilarte ML, Tymczyszyn EE, Serradell M de los Á, Carasi P. Freeze-drying of *Enterococcus durans*: Effect on their probiotics and biopreservative properties. LWT - Food Science and Technology, 2021; **137**, 110496

[69] Pieniz S, Andreazza R, Anghinoni T, Camargo F, Brandelli A. Probiotic potential, antimicrobial and antioxidant activities of *Enterococcus durans* strain LAB18s. Food Control. 2014; **37**: 251-256

[70] Amaral DMF, Silva LF, Casarotti SN, Nascimento LCS, Penna ALB. *Enterococcus faecium* and *Enterococcus durans* isolated from cheese:

Survival in the presence of medications under simulated gastrointestinal conditions and adhesion properties. J. Dairy Sci. 2017; **100**:933-949

[71] Haubert L, dos Santos Cruxen CE, Fiorentini AM, da Silva WP. Tetracycline resistance transfer from foodborne *Listeria monocytogenes* to *Enterococcus faecalis* in Minas Frescal cheese. International Dairy Journal. 2018; **87**:11-15

[72] Cocconcelli PS, Cattivellia D, Gazzola S. Gene transfer of vancomycin and tetracycline resistances among *Enterococcus faecalis* during cheese and sausage fermentations. International Journal of Food Microbiology. 2003; **88**:315– 323

[73] Dat NM, Hamanaka D, Tanaka F, Uchino T. Control of milk pH reduces biofilm formation of *Bacillus licheniformis* and *Lactobacillus paracasei* on stainless steel. Food Control. 2o12; **23**:215-220

[74] I.C. Fernández-No, M. Guarddon, K. Böhme, A. Cepeda, P. Calo-Mata, J. Barros-Velázquez*, Detection and quantification of spoilage and pathogenic *Bacillus cereus*, Bacillus subtilis and Bacillus licheniformis by real-time PCR, Food Microbiology 28 (2011) 605-610

[75] Soares MB, Almada CN, Almada CN, Rafael CR, Martinez RCR, Pereira EPR, Balthazar CF, Adriano G. Cruz AG, Ranadheera CS, Sant'Ana AS. Behavior of different *Bacillus* strains with claimed probiotic properties throughout processed cheese ("requeijão cremoso") manufacturing and storage. International Journal of Food Microbiology. 2019; **307**, 108288

[76] Pierron A, Zayet S, Toko L, Royer PY, Garnier P, Gendrin V.' Catheter-related bacteremia with endocarditis caused by *Kocuria rhizophila*, Médecine et maladies infectieuses, 2020, DOI: 10.1016/j.medmal.2020.09.007.

[77] Osuka H, Hitomi S, Koganemaru H, Kaneko T. A case of bacteremia caused by *Hafnia paralvei*, J Infect Chemother. 2011; **17**:855-857.

Chapter 7

Fermentation of Cocoa Beans

Romel E. Guzmán-Alvarez and José G. Márquez-Ramos

Abstract

Cocoa bean fermentation is a spontaneous process driven by an ordered microbial succession of a wide range of yeasts, lactic acid and acetic acid bacteria, some aerobic sporeforming bacteria and various species of filamentous fungi. The process of cocoa fermentation is a very important step for developing chocolate flavor precursors which are attributable to the metabolism of succession microbial. The microbial ecology of cocoa has been studied in much of the world. In Venezuela, studies have been carried out with Criollo, Forastero, and Trinitario cocoa, fermented under various conditions, the results obtained coinciding with the reported scientific information. Fermentation must be associated with the type of cocoa available, carried out knowing the final processing and derivative (paste, butter, powder). The results shown in this chapter correspond to investigations carried out with cocoa from three locations in Venezuela. The quantification, identification, isolation, functionality of the most representative microbiota involved in the fermentation of these grains was sought. This to give possible answers to the fermentation times and improvement of the commercial quality. Likewise, generate greater interest on the part of the producers in carrying out the fermentation.

Keywords: cocoa beans, fermentation, microbial ecology

1. Introduction

Cocoa beans (*Theobroma cacao* L) are the basis for the production of cocoa powder and butter, as well as chocolate. In order to obtain fine cocoa beans of aroma it is necessary to carry out a process of benefit (harvesting, shelling, fermentation, drying) in which the spontaneous fermentation of the grains has great influence, since in addition to an interesting occurrence Microbial succession favors the production of enzymatic, chemical, physicochemical reactions among others that still need to be studied.

The part used of the cacao tree *Theobroma cacao* L. are the beans and the edible is its cotyledons, which undergo important changes during the fermentation, drying and manufacturing process, giving rise to a flavor and aroma that is appreciated by chocolate consumers around the world. The raw cocoa has an extremely bitter and astringent taste, so it must be treated by a process in which the microorganisms, through fermentation, modify their components. The fermentation of the cocoa begins with the opening of the ear and the extraction of the cocoa beans. The grains that generally (depending on the maturity and time of opening) at the moment of being extracted are wrapped in a white mucilaginous pulp that comprises 40% of the wet basis weight of the fresh grain. Composed mainly of 82–87% water, 10–15% fermentable sugars (glucose, fructose and sucrose), 2–3% pentoses, 1–3% citric acid, 1–1.5% pectin, in addition to 1–2% of other hemicellulosic polysaccharides.

Proteins, amino acids, vitamins (mainly vitamin C) and minerals are also present, the mucilaginous pulp is an excellent substrate for microbial growth [1–3]. The process of fermentation of the cocoa is natural, since usually they are not intentionally added initiating microorganisms to the grains, which are sterile inside the ears and are contaminated with different microorganisms coming from all the surfaces with which they come into contact (tear, knives and the hands of the people who manipulate it).

The fermentation process of cocoa is characterized by a microbial succession in which according to various authors [3–9], the yeasts participate first, then the lactic acid bacteria (LAB) act and, finally, the acetic bacteria (AAB) intervene. Additionally, spore-forming bacilli of the genus Bacillus, and filamentous fungi. This process is essential both to modify the beans, eliminating the mucilage, and to prepare the grain that requires a battery of enzymes responsible for modifying its color, taste and smell.

They have been reported [3–6, 10–12] more than 100 microbial species that show different metabolic properties, identifying themselves new species due to the improvement in cultivation techniques together with the use of molecular biology tools. Two dominant bacterial species *Lactobacillus fermentum* and *Acetobacter pasteurianus*, together with four yeast species *Saccharomyces cerevisiae*, *Hanseniospora thailandica*, *H. opuntiae* and *Pichia kudriavzevii*, represent the central component of the bacterium-fungus association that lead to the fermentation of cocoa in many of the regions where it is produced [13].

The term cocobiota, was introduced [13] to refer to the microbial association of bacteria and fungi involved in the spontaneous fermentation of cocoa beans, which originate metabolites present in cocoa powder and dark chocolate, which can have beneficial effects on health. Five main groups of microorganisms participate in cocoa fermentation: filamentous fungi, yeasts, lactic acid bacteria, acetic acid bacteria, and various Bacillus species [5].

Venezuela has specimens of cacao (*Theobroma cacao* L) in almost all its territory; however, the study of the microbial characterization developed during fermentation has been very little studied compared to the reports found today worldwide. Therefore, the present investigation was based on quantifying, identifying, isolating, and knowing the functionality of the most representative microorganisms involved in the fermentation of cocoa beans grown in three locations, where the work systems were completely different.

In most producing countries, including Venezuela, cocoa fermentation is a process that is carried out in an artisanal way, using systems such as plastic baskets, wooden drawers, and staggered wooden boxes. These systems are generally covered with leaves of Musaceae or heliconia. Specifically, in Venezuela, some localities use a fermentation system in plastic containers for a time that is between 12 and 24 hours. This fermentation system in Venezuela has allowed women's inclusion in the workplace, thereby achieving gender equality that has caused so many problems worldwide. It is fulfilled with the fifth (5th) Goal to Transform Our World proposed by the United Nations for its 2015–2030 plan.

2. Cocoa fermentation

According to the fermenter used, cocoa beans' fermentation is done in quantities of 25 kg up to groups of approximately 200–250 kg—the type of cocoa influences the time and the technique used. Forastero and Trinitario cocoa require between 5 to 7 days of fermentation, while Criollo requires 2 to 3 days. After seven days of fermentation, some off-flavors may be formed by some fungal species. During fermentation, temperature changes occur that favor both the evolution of microorganisms and a set of enzymes [14]. The changes result in the formation of cocoa aroma precursors (amino acids and reducing sugars). Precursors are enhanced in the stages of drying and roasting.

The fermentation process is characterized by a well-known microbial succession. The initial pH (3.6) of the pulp caused by the presence of citric acid, together with low oxygen levels, favors yeasts' colonization. The proliferation of these leads to the production of ethanol and the secretion of pectinolytic enzymes.

2.1 Yeast

The yeast population increases in number within the first 24 h of fermentation, after which it slowly decreases. Studies in cocoa fermentation from different areas of Venezuela show that the behavior of the yeasts (**Table 1**) occurs in a similar way when it has been fermented for three days for Criollo cocoa (from Sur del Lago, Zulia state) and five to seven days for Forastero cocoa (from Barlovento, Miranda state).

Table 1 shows the log CFU/g of the yeasts found in the fermented cocoa beans in the different sampled sites. The yeast population was present from the beginning of fermentation in all samples, with counts from 1.00 LogUFC/g for replica 2 (B^2) from the INIA Tapipa experimental station, Miranda state, to 5.31 LogUFC/g for replica 3 (C^3) of the INIA Chama experimental station, Zulia state. The yeasts' behavior was variable, finding maximum counts of 1.69 LogUFC/g for the first day of fermentation (C^1) up to 1.00 LogUFC/g (C^1 and C^2). Likewise, for the Miranda state's replicas, various behaviors were observed, which experienced increases or decreases according to the days of fermentation.

According [15], there is an increase in the number of yeasts by 10^3 CFU/g around the sixth to the seventh day of fermentation. This is possibly due to the growth of thermotolerant yeasts that use some of the acids, coinciding with an increase in oxygen content in the mass and survivors in the fermentation system's colder external zones. The experimental stations in Zulia showed initial values of 5.31 Log CFU/g and a significant decrease in the population at the end of the process. The variations found in the yeast population count for the studied sites could probably be due to deviations inherent in each locality's fermentation process. Genetic variations of the plantations, fermentation methods; plastic baskets (used in the

	Fermentation time (Days)					
Sample	**0**	**1**	**2**	**3**	**4**	**5**
A	3.48 ± 0.03^c	3.27 ± 0.23^a	1.67 ± 0.06^a	1.00 ± 0.01^a	4.17 ± 0.17^a	3.61 ± 0.08^a
B^1	4.42 ± 0.21^b	1.00 ± 0.01^b	1.80 ± 0.07^a	1.83 ± 0.07^a	4.55 ± 0.30^a	1.67 ± 0.15^b
B^2	1.00 ± 0.01^e	1.69 ± 0.04^b	1.73 ± 0.05^a	1.74 ± 0.06^a	3.21 ± 0.12^b	4.33 ± 0.10^a
C^1	2.49 ± 0.17^d	1.69 ± 0.07^b	1.00 ± 0.01^a	1.00 ± 0.01^a		
C^2	2.76 ± 0.09^d	1.33 ± 0.05^b	1.00 ± 0.01^a	1.00 ± 0.01^a		
C^3	5.31 ± 0.06^a	1.00 ± 0.01^b	1.00 ± 0.01^a	1.35 ± 0.03^a		

^{a,b,c,d,e}Different letters for each row indicate significant differences ($p < 0,05$).
Mean ± standard deviation.
A: La Trinidad fermented cocoa, Miranda state B^1 and B^2: INIA Tapipa fermented cocoa, Miranda state C^1, C^2, C^3: INIA Chama fermented cocoa, Zulia state. 1,2,3: indicates the replica number.

Table 1.
Log CFU/g of yeasts during the fermentation of cocoa beans from one location and five experimental stations.

fermentation of cocoa beans from the Miranda experimental stations) or wooden crates (used in the fermentation of cocoa beans from the Zulia experimental stations), failures in the control of the artisanal process, fermentation incomplete, excessive turning of the grains, changes in oxygen content, pH, accumulation of ethanol and other compounds can affect the fermentation system.

Studying the chemical and microbiological composition [16] during the fermentation process of cocoa beans from the Chuao area in Aragua state, found a yeast count for the first days of fermentation in the order of 3.77 to 5.38 Log CFU/g, while [17], in Venezuelan cocoa beans of the Carenero variety, grown in Merecure, Miranda State, reports yeast populations of 8.21 Log CFU/g, at week 1 of fermentation reaching a population density of 6.80 CFU/g at week 3 fermentation, decreasing its number to 3.00 Log CFU/g at the end of the fermentation process. The yeasts are predominant at 12 and 36 hours after the start of the fermentation of cocoa beans [3, 17].

In fermented cocoa beans in the Southeast region of Ivory Coast [17], found a rapid increase in the yeast population, going from 10^6 CFU/g to 10^7 CFU/g at 36 h, after 84 h, this decreased to 10^1 CFU/g. Studying [18] the diversity of yeasts involved in the fermentation of cocoa from six of the principal producing regions of Ivory Coast founded a yeast population of 10^4–10^5 CFU/g up to 10^7–10^8 CFU/g between 12 and 24 hours of fermentation. Growth kinetics were very similar for the six regions, with an increase in the yeast population during the first 24 hours of fermentation and a progressive decrease after 24 hours after starting the fermentation process.

Table 2 shows the macroscopic morphological characteristics of fourteen (14) yeast strains isolated at the sampled sites. 4 isolated from the cocoa beans fermented in $A^{1,2}$ from the beans fermented in B^1, 2 from B^2, 2 from C^1, C^2, and C^3. In most isolates, the morphology found agrees with the described by [19].

Strain	Shape	Margin	Elevation	Surface	Color	Brightness	Texture
A	Oval	Curly	Umbilicada	Rough	Cream	Opaque	Butyrose
A	Oval	Curly	Umbilicada	Rough	Cream	Opaque	Butyrose
A	Ovoid	Whole	Flat	Smooth	Cream	Brilliant	Butyrose
A	Elliptical	Whole	Flat	Smooth	Cream	Opaque	Butyrose
B^1	Ovala	Curly	Umbilicada	Rough	Cream	Opaque	Butyrose
B^1	Elliptical	Whole	Flat	Smooth	Cream	Opaque	Butyrose
B^2	Oval	Curly	Umbilicada	Rough	Cream	Opaque	Butyrose
B^2	Oval	Curly	Umbilicada	Rough	Cream	Opaque	Butyrose
C^1	Spherical	Whole	Convex	Smooth	Pink	Brilliant	Butyrose
C^1	Round	Rhizoid	Umbilicada	Smooth	Cream	Opaque	Butyrose
C^2	Oval	Curly	Umbilicada	Rough	Crema	Opaque	Butyrose
C^2	Oval	Rhizoid	Umbilicada	Smooth	White	Opaque	Butyrose
C^3	Spherical	Whole	Convex	Smooth	Pink	Brilliant	Butyrose
C^3	Oval	Curly	Umbilicada	Rough	Cream	Opaque	Butyrose

A: La Trinidad fermented cocoa, Miranda state B^1 and B^2: INIA Tapipa fermented cocoa, Miranda state C^1, C^2, C^3: INIA Chama fermented cocoa, Zulia state. 1,2,3: indicates the replica number.

Table 2.
Morphological characteristics of yeasts isolated from fermented cocoa beans in one location and five experimental stations.

At a macroscopic level, we could observe six (6) different phenotypic of yeast in the isolated colonies' morphology. Most (8) colonies with oval shape, wavy margin, umbilicated, cream-colored, rough surface, opaque, and butyric texture were found in all the studied sites. Specifically, from sample A, there was (1) one ovoid colony, whole, flat, smooth cream-colored isolated and another colony (1) with an elliptical shape, a whole margin, flat, smooth, shiny, cream-colored, and butirose. For samples C^1 and C^3. Two (2) isolates with a spherical shape, entire margin, convex, smooth, shiny, and pinkish color. One (1) colony with a round shape, with rhizoid margin, umbilicated, smooth, opaque, cream-colored for sample C^2. Finally, one isolated yeast colony of sample C^3 oval, with rhizoid margin, smooth, opaque. Existence of at least six different yeast genera in the fermented cocoa beans [19]. The isolates identified would correspond to the genus *Saccharomyces*, *Debaryomyces*, *Hanseniospora*, *Rhodotorula*, *Pichia*, and *Candida*.

However, to confirm this, more biochemical tests must be performed and correlated with molecular tests. All these genera and many others have been found in cocoa beans' fermentation from different countries [4, 6, 10, 20–24]. The macroscopic characteristics found in the isolates are due to the fact that during cocoa fermentation an ecological succession occurs with micro-environmental changes such as the availability of nutrients, changes in pH, temperature, oxygen, among others. The yeasts are capable of changing their characteristics and behavior, according to the environmental conditions in the substrate that they are [25–27].

In relation to the physiological identification of the 6 groups of yeasts isolated from the fermented cocoa beans in the studied sites, **Table 3** shows the results obtained for the sugar fermentation. Glucose positively ferments for all genera with the exception of the strain identified as L2, which fermented it slowly and weakly, and the L4 strain gave a negative result for both glucose fermentation and the other seven sugars used in the test Likewise, the strain identified as L1 was the only one that showed positive fermentation for maltose, galactose and sucrose and weakly for raffinose, cellobiose was only slowly fermented by L3, while strains L5 and L6 only managed to ferment positively glucose. Lactose and xylose could not be fermented by any of the 6 groups of yeasts isolated and characterized in this study.

Investigating yeasts' diversity and role in the spontaneous fermentation of cocoa beans from Indonesia [22]. Isolated seven yeast strains that they identified based on similar phenotypic characteristics with eight reference strain *Saccharomyces cerevisiae*, *Saccharomyces bayanus, Candida tropicalis, Candida krusei, Candida lambica, Saccharomycopsis fibuligera, Saccharomycopsis fermentant*, and *Kloeckera* sp. These eight reference strains were selected based on the similarity in phenotypic characteristics with the seven isolates and that they are the yeasts most frequently

Strain	Glucose	D-Xilose	Maltose	Galactose	Lactose	Sacarose	Raffinose	Cellobiose
L1	+	—	+	+	—	+	W	—
L2	S	—	—	L	—	L	S	—
L3	+	—	—	—	—	S	—	+/−
L4	—	—	—	—	—	—	—	—
L5	+	—	—	—	—	—	—	—
L6	+	—	—	—	—	—	—	—

+: positive, L: positive delayed, S: slow positive, W: weak positive, −: negative.

Table 3.
Fermentation properties of eight (8) sugars from 6 yeasts isolated from cocoa beans fermented in one locations of Venezuela.

found in fermented cocoa. The carbon sources preferred by most yeasts are hexoses such as glucose and fructose, present in high concentrations in cocoa beans, and mannose since they can enter the glycolytic pathway directly. Before entering this metabolic pathway, the first and limiting step for glucose metabolism is its transport through the membrane. The consumption of this sugar is mediated by a large family of proteins associated with carbon sources [26].

The L1 strain was the only one that positively fermented sucrose. It has been shown that this sugar is converted to glucose and fructose in species such as *S. cerevisiae* for its use; this event takes place in the yeast cell membrane thanks to periplasmic invertase. The invertase gene expression (SUC2) is repressed when there are high concentrations of glucose and reactivated when the concentration of this sugar falls [23]. The cocoa bean pulp contains approximately 14% sugars (60% sucrose and 39% a mixture of glucose and fructose). Assimilation and fermentation with the efficiency shown are because the yeasts identified as L1 show a certain degree of specialization to ferment this substrate thanks to the proteins mentioned above and enzymes' presence.

Sugars such as maltose, galactose, cellobiose, and raffinose could also be used by these yeasts, although to a lesser extent than glucose and sucrose. Could tentatively identify This group of isolates identified as L1 as belonging to the species *S. cerevisiae*. This species of yeast isolated as one of the most dominant species in the fermentation of cocoa beans from several countries such as Belize, Brazil, Ivory Coast, Ecuador, Ghana, Indonesia, Mexico, Nigeria, Peru, Nicaragua, and the Dominican Republic [2, 6, 9, 10, 20–23, 27–33] due to that there is evidence that it is a fast-growing yeast and that it presents high tolerance to ethanol thanks to the presence of fatty acids contained in the cell and also because it presents a vigorous pectinase and endo and exo protease activity, which influences the quantity of free amino acids in the ferments contributing beneficially to the flavor of the cocoa beans [34, 35].

The isolates identified as L2 were the only ones that fermented glucose slowly after 14 days, as did galactose and sucrose slowly. These could be the *Debaryomyces hansenii* species. Yeasts "oilseeds" that accumulate lipids. It is a highly heterogeneous species, and therefore versatile, as evidenced by phenotypic differences between strains, such as variations in its ability to assimilate and ferment various carbon sources, the expression of different lipase and protease activities, and its diversity under conditions growth [36]. Although this species of yeast is not common in cocoa fermentation, it is used in starter cultures for the fermentation of cocoa beans [29, 37].

Cellobiose was fermented only by the L3 strain. Depending on the species, different strategies have been observed among yeasts to use cellobiose. One of them consists of the expression of extracellular β-glucosidases on the cell surface; cellobiose is hydrolyzed extracellularly. After this, glucose is transported and metabolized inside the cell. The second strategy consists of the phosphorolytic pathway; cellobiose is transported into the cell by cellodextrins. An intracellular phosphorylase cuts the disaccharide with an inorganic phosphate, producing a glucose molecule and an α-glucose-1-phosphate that can be metabolized quickly. The hydrolytic pathway is the third strategy. Sugar is transported into the cell by cellodextrins to later be hydrolyzed by a β-glucosidase into two glucose molecules that can be easily metabolized by the cell [23].

According to [19], these isolates could be identified as *Hanseniospora opuntiae*. This species is considered one of the most frequent in cocoa fermentation processes [9, 38–41]. For example, it has been described as an elemental microbiota species in Malaysia's cocoa fermentation processes [42]. In general, this is a persistent species throughout the fermentation process, regardless of the cocoa's origin and the method used to ferment the beans. A study with cocoa from Ecuador, both in boxes and in fermentation platforms, exhibited the presence of *Hanseniaspora* sp. (100% identity with *H. opuntiae*, *H. guilliermondii* and *H. uvarum*) and *P. kudriavzevii* [43]

The isolates grouped as L4 did not show fermentation for any of the eight sugars used; however, they could grow in the tubes. According to [19], several species of yeasts grow aerobically on sugars that fail to ferment. From the pattern found for this isolate, L4 could be classified as *Rhodotorula* spp. This species has not been reported in previous studies [23] on cocoa fermentation, identified *R. dairensis* in fermented cocoa beans in two locations in Ecuador.

L5 and L6 showed a very similar fermentation pattern for sugars used, which allows us to presume that they could be the same species, *Pichia kudriavzevii* or *Candida tropicalis*. The characterization by molecular biology will enable us to distinguish between which of the two species belong. Both species are recurrent in the cocoa fermentation processes [22, 39, 43–45]. The species *P. kudriavzevii* is also part of the central microbiota of cocoa ferments from Malaysia. In addition to his thermotolerance, its presence is favored by tolerance to highly selective environmental conditions present in the fermentation of cocoa pulp, such as the presence of chemical stressors, such as ethanol, acetic acid, and lactic acid [42, 46].

None of the isolated and identified species could ferment xylose; this is because the yeasts involved in cocoa ferments cannot operate the proton-symport system, subject to catabolic repression and the facilitated diffusion system for the transport and use of this sugar. Neither species could ferment lactose; this sugar [47] can be assimilated or fermented by yeasts through the transport via induction of a lactose-permease and its subsequent hydrolyzation to glucose and galactose by an intracellular β-glucosidase.

2.2 Lactic Acid Bacteria (LAB)

The values of the Log CFU/g of the LAB are shown in **Table 4**. A low population was obtained from 1.00 CFU/g to 2.75 CFU/g at the beginning of the fermentation process. LAB increased after two days and decreased in the last days of the cocoa beans' fermentation process. LAB population for the cocoa samples from the Zulia region (C^1-C^3) was 2.00 Log CFU/g, a logarithmic cycle above the representatives from the Miranda region (A and B^1). B2 was in the same order of C^1-C^3.

The LAB growth dynamics found for the regions of Miranda and Zulia could be due to several factors. The frequency of turning or removal of the mass of grains in fermentation, a process that It involves aeration or ventilation of the cocoa mass, generating exposure to oxygen and an increase in its concentration in the cocoa mass; in this investigation, this process was carried out every 24 hours. This fact could cause a significant reduction in the number of the LAB population during the last days of the fermentation process. In general, LAB are facultative anaerobic microorganisms. Their optimal growth conditions are in oxygen-free atmospheres, but they can tolerate low concentrations; therefore, oxygen is one of the various factors that generate microbial "stress" in this group. Similar results have been reported for different fermentation processes of cocoa beans from different regions of the world [21, 48].

In cocoa beans' fermentation, LABs exhibit the fastest growth rate during the 16–48 hours of fermentation and are present in large numbers, but not necessarily in biomass than yeasts for a short period [15].

The preliminary identification of LAB (**Table 5**) was selected colonies with bacillary or rounded shape, creamy white or beige color, positive Gram stain, catalase/oxidase negative. According to the results, there were four (4) different BAL phenotypes within the isolated strains. Presented Most of them (12) as long bacilli

Sample	Fermentation time (Days)					
	0	1	2	3	4	5
A	1.00 ± 0.01^{c}	2.79 ± 0.13^{ab}	3.27 ± 0.08^{ab}	2.92 ± 0.06^{a}	2.44 ± 0.16^{a}	1.33 ± 0.05^{a}
B^1	1.53 ± 0.09^{bc}	2.45 ± 0.23^{b}	2.48 ± 0.19^{cd}	2.49 ± 0.21^{ab}	1.99 ± 0.08^{a}	1.73 ± 0.06^{a}
B^2	1.87 ± 0.08^{abc}	1.63 ± 0.12^{c}	2.36 ± 0.14^{cd}	2.49 ± 0.17^{ab}	1.58 ± 0.13^{a}	1.33 ± 0.05^{a}
C^1	2.27 ± 0.15^{ab}	3.22 ± 0.21^{a}	1.68 ± 0.04^{b}	3.75 ± 0.12^{a}		
C^2	2.75 ± 0.08^{a}	1.73 ± 0.15^{c}	2.19 ± 0.087^{d}	2.50 ± 0.04^{ab}		
C^3	2.45 ± 0.17^{ab}	2.75 ± 0.03^{ab}	2.88 ± 0.06^{bc}	2.32 ± 0.09^{ab}		

^{a,b,c,d,e}*Different letters for each row indicate significant differences ($p < 0,05$). Mean ± standard deviation.*
A: La Trinidad fermented cocoa, Miranda state B^1 and B^2: INIA Tapipa fermented cocoa, Miranda state C^1, C^2, C^3: INIA Chama fermented cocoa, Zulia state. 1,2,3: indicates the replica number.

Table 4.
Log CFU/g of LAB during the fermentation of cocoa beans.

	Morphology/Gram positive	Catalase	Oxidase	Glucose Gas	15°C	47°C
BAL1-A	Short bacillus	—	—	+	+	—
BAL2-A	Short bacillus	—	—	+	+	—
BAL3-A	Long bacillus	—	—	+	+	—
BAL4-A	Long bacillus	—	—	+	+	+
BAL1-B[1]	Short bacillus	—	—	+	+	+
BAL2-B[1]	Coccobacilli	—	—	—	—	—
BAL3-B[1]	Long bacillus	—	—	+	+	—
BAL4-B[1]	Long bacillus	—	—	+	+	—
BAL5-B[1]	Short bacillus	—	—	+	+	—
BAL1-B[2]	Long bacillus	—	—	+	+	—
BAL2-B[2]	Cocci	—	—	—	+	—
BAL3-B[2]	Long bacillus	—	—	+	+	—
BAL4-B[2]	Bacilluscocco	—	—	+	+	—
BAL5-B[2]	Cocci	—	—	—	+	—
BAL1-C[1]	Long bacillus	—	—	+	+	—
BAL2-C[1]	Long bacillus	—	—	+	+	—
BAL3-C[1]	Short bacillus	—	—	+	+	—
BAL1-C[2]	Long bacillus	—	—	+	+	—
BAL2-C[2]	Long bacillus	—	—	+	+	—
BAL3-C[2]	Short bacillus	—	—	+	+	+
BAL1-C[3]	Short bacillus	—	—	+	+	—
BAL2-C[3]	Long bacillus	—	—	+	+	—
BAL3-C[3]	Short bacillus	—	—	+	+	+
BAL4-C[3]	Long bacillus	—	—	+	+	—

+: Growth, –: no growth.

Table 5.
Morphological and biochemical characteristics of BAL isolated from fermented cocoa beans.

(BAL1), eight (8) isolated were short bacilli (BAL2), two (2) in the shape of coccobacilli (BAL3), and in the form of cocci, only three (3) isolated (BAL4). Similar results have been reported for different fermentation processes of cocoa beans from different regions of the world [7, 18, 21, 27, 30, 49].

According to these results, we could classify 22 isolates as Lactobacilli, belonging to the genus Lactobacillus and three (3) as Lactococci, genus Lactococcus. Twenty-one (21) of the isolates showed growth at 15° C, while only four (4) could grow at 15 and 47° C. Most of the twenty-two isolates (22) produced gas from glucose fermentation. Accordingly, the former could be heterofermenting LAB and the latter homofermenting LAB.

2.3 Acetic Acid Bacteria (AAB)

The quantification of AAB (**Table 6**) shows that populations between 2.32–3.62 Log CFU/g and 1.33–5.31 Log CFU/g were obtained in the first days of fermentation for samples from the Miranda and Zulia experimental stations, respectively.

Sample	Fermentation time (Days)					
	0	1	2	3	4	5
A	2.32 ± 0.14^{c}	2.25 ± 0.22^{d}	3.37 ± 0.20^{b}	3.41 ± 0.12^{b}	4.75 ± 0.12^{a}	5.27 ± 0.20^{a}
B^{1}	3.62 ± 0.24^{b}	1.73 ± 0.06^{d}	1.00 ± 0.01^{d}	1.76 ± 0.09^{c}	1.86 ± 0.074^{b}	5.44 ± 0.16^{a}
B^{2}	2.52 ± 0.16^{c}	6.18 ± 0.18^{a}	6.36 ± 0.07^{a}	6.04 ± 0.06^{a}	1.33 ± 0.05^{b}	1.00 ± 0.01^{b}
C^{1}	2.18 ± 0.18^{c}	3.43 ± 0.16^{c}	2.68 ± 0.21^{c}	3.60 ± 0.01^{b}		
C^{2}	1.33 ± 0.06^{d}	4.39 ± 0.09^{b}	3.19 ± 0.19^{b}	3.31 ± 0.14^{b}		
C^{3}	5.31 ± 0.20^{a}	4.26 ± 0.22^{bc}	3.56 ± 0.08^{b}	2.42 ± 0.11^{b}		

^{a,b,c,d,e}Different letters for each row indicate significant differences ($p < 0,05$). Mean ± standard deviation.
A: La Trinidad fermented cocoa, Miranda state B^{1} and B^{2}: INIA Tapipa fermented cocoa, Miranda state C^{1}, C^{2}, C^{3}: INIA Chama fermented cocoa, Zulia state. 1,2,3: indicates the replica number.

Table 6.
Log CFU/g of AAB during the fermentation of cocoa beans.

AAB carry out the transformation of the ethanol produced by the yeasts into acetic acid. The conversion of ethanol into acetic acid is an exothermic reaction. Ethanol and acetic acid diffuse into the cocoa beans, which finally generates the embryo's impossibility to develop a new cocoa plant.

The macroscopic, physiological, and biochemical morphological characteristics of twenty (20) BAA isolated from the fermented cocoa beans are shown in **Table 7**. Five (5) strains were isolated from A, two (2) from B1, four (4) from B2, three (3) from C1, two (2) from C2, and four (4) from C3. 100% of the isolates were catalase positive, oxidase negative, and capable of oxidizing acetate, a characteristic behavior of Acetobacter species. Except for A. peroxydans [50].

Two (2) morphotypes were found. Bacilli (BAA1), colonies with a point shape, 5 mm in diameter, convex, creamy beige, and coccobacilli (BAA2) elongated colonies. 2 mm, connected, shiny, beige; both Gram-negative, catalase-positive, and oxidase negative. From the above, there are four (4) subtypes. Eight (8) isolates cannot grow in 10% ethanol and 30% D-glucose; identified as BAc1, they could be the species *A. syzygii*, *A. lovaniensis*, *A. pomorum*, *A. aceti*, or *A. tropicalis*. Seven (7) isolates did not grow in 10% ethanol, but in 30% D-glucose classified as BAc2, presumably *A. ghanensis* or *A. cibinongensis*. Two (2) isolates grown in 10% ethanol (BAc3), *A. oeni*, or *A. pasteurianus*. Three (3) that grow in 10% ethanol and 30% D-glucose called BAc4, possibly they are the species *A. nitrogenifigens* [50] (**Table 7**).

	Morphology/ Gram negative	Catalase	Oxidase	10% Ethanol Growth	30% D-glucose Growth	Acetate oxidation
BAA1-A	Bacillus	+	—	—	—	+
BAA2-A	Bacillus	+	—	—	+	+
BAA3-A	Coccobacilli	+	—	—	—	+
BAA4-A	Bacillus	+	—	+	—	+
BAA5-A	Bacillus	+	—	—	—	+
BAA1-B[1]	Bacillus	+	—	+	+	+
BAA2-B[1]	Bacillus	+	—	—	+	+
BAA1-B[2]	Bacillus	+	—	—	+	+
BAA2-B[2]	Bacillus	+	—	+	+	+
BAA3-B[2]	Bacillus	+	—	—	+	+
BAA4-B[2]	Bacillus	+	—	—	—	+
BAA1-C[1]	Bacillus	+	—	—	+	+
BAA2-C[1]	Bacillus	+	—	—	—	+
BAA3-C[1]	Bacillus	+	—	—	—	+
BAA1-C[2]	Coccobacilli	+	—	—	—	+
BAA2-C[2]	Bacillus	+	—	—	—	+
BAA1-C[3]	Bacillus	+	—	—	+	+
BAA2-C[3]	Bacillus	+	—	—	+	+
BAA3-C[3]	Bacillus	+	—	+	+	+
BAA4-C[3]	Bacillus	+	—	+	—	+

+: Growth, −: no growth.

Table 7.
Morphological and biochemical characteristics of BAL isolated from fermented cocoa beans.

3. Evaluation of fermented cocoa

Once the cocoa is fermented, typing analysis is done to classify it. The cut test (**Figure 1**) is to identify cocoa beans that have been well-fermented, lightly fermented, and non-fermented. However, it will allow you to observe the grains that present problems, including; over-fermented, slaty, sprouts, affected by insects.

Figure 1 schematizes the portion of a cutting test (carried out manually with the help of a knife) of cocoa beans. This test seeks to know the percentage of fermentation that specific cocoa has—the determination to be made in at least five (5) replicas of 100 grains each.

Cocoa is classified commercially according to the percentage of fermentation. In Venezuela, the established classification [51, 52] in Extra Fine (EF: 95%), Fine First (F1: 80%), Fine Second (F2: 20%). It is known as ordinary cocoa beans with fermentation percentages lower than F2.

It is an international consensus that cocoa beans with brown coloration and well-defined streaks on their cotyledons are fully fermented. Grains with violet/brown colorations are associated with slightly fermented grains. Grains purplish, compact, or very little change in the violet color (typical of Trinitarios and Forasteros) may be unfermented. **Figures 2–4** show cocoa beans with these characteristics.

In **Figure 2**, cocoa beans with violet coloration are shown, indicating that they are not fermented. In the dissection of these beans, it is observed that there are no well-defined cracks or channels, which are formed due to the microbiological succession and enzymatic reactions that occur during the fermentation process. The bean shown in the upper left part has a dark, blackish coloration, which indicates that it is a slate grain, not desirable in transforming from cocoa to chocolate. The dissection of the lower-left grain shows a violet coloration and compaction in it, indicative of non-fermented grains; it also presents white areas that could indicate the presence of fungi or attack by insects.

Figure 1.
Cutting test of cocoa beans from various cocoa-producing areas of Venezuela, representing fermented, slightly fermented, unfermented, and damaged beans.

Figure 2.
Unfermented cocoa beans (upper and lower left) and well-fermented cocoa beans (upper and lower right).

Figure 3.
Slate cocoa bean with insect attack.

The upper-right grain has a brown color and a notable presence of cracks or channels in its cotyledons that indicate that it has completed the fermentation process. Another indication of a well-fermented grain is observed in the lower-right grain, in which, although cracks and channels are not perfectly evident, it presents the separation or opening of its cotyledons, which allows classifying them as well or completely fermented.

Slate cocoa beans (**Figure 3**) generally appear due to a lack of fermentation or incomplete fermentation (which is interrupted by the producer for various reasons). Slate grains are associated with increased bitterness and astringency. Attributes that later affect the cocoa paste (liquor) and, therefore, the production of chocolate.

Figure 4.
Slightly or partially fermented cocoa beans.

This bean also has a compact appearance (not fermented). The bean shown a cavity in the center indicates that insects have attacked it. Generally, this occurs because there are poor storage conditions (relative humidity and temperature). Proliferating species such as *Ephestia elutella*, which in addition to causing damage to the beans, can affect the production of chocolate.

Figure 4 shows cocoa beans that have passed the fermentation process; however, it has not been complete for them. Obtaining slightly or partially fermented grains occurs because the fermentation mass is excessive; there are no controlled processes for removing the fermentation mass or due to the location of the grains in the drawers or fermentation systems, in addition to the amount of mucilage that each grain has. In this image (**Figure 4**), the dissection of the upper-right grain is an excellent example of partially fermented. The periphery of the grain is purple, and the center of the grain is brown, with the appearance of channels.

Forasteros-type grains, as they present a natural intense violet color, can often pass as unfermented or partially fermented. Therefore, it is necessary to know the evaluation criteria very well to avoid incurring classification errors.

Finally, it is essential to note that the purpose of the cutting test is to establish a commercial classification and does not establish a precise and exact indicator of what the aromatic or flavor quality of the cocoa beans will be.

Another parameter to consider in fermented cocoa is the percentage of acidity because many times high fermentation percentages (greater than 90%) generate a higher content of acetic acid (predominant acid in cocoa due to the action of acetic acid bacteria). The high acidity makes cocoa paste/liquor require long working times in chocolate conching machines. In the case of Venezuela, the relationship between cocoa producer-chocolatier-artisan's increases day by day. With this, criteria can be established regarding the final fermentation percentages that are desired for a specific type of cocoa.

4. Conclusion

Fermentation is a crucial stage in the post-harvest processing of cocoa. The succession of yeasts, lactic acid bacteria, acetic acid bacteria, and others more involved, together with the system's conditions and the enzymes, allows the formation of aroma precursors (sugars and amino acids). Finally, the precursors are magnified in the roasting, resulting in a diversity of aromas in the final derived products.

It is necessary to deepen each type of cocoa fermentation to know and standardize the percentage or degree of fermentation that each bean requires. It is important to note that 100% fermented grains often provide an acetic acid index that ultimately translates into longer working time in chocolate conching machines.

Acknowledgements

Researchers thank the staff of the Tapia and Chama experimental stations of the National Agricultural Research Institute (INIA) for their valuable collaboration in obtaining cocoa specimens. We thank the producers of the state of Miranda for being involved in developing the fermentation study.

Conflict of interest

The authors whose names are listed in this paper certify that they have NO affiliations with or involvement in any organization or entity with any financial interest (such as honoraria; educational grants; participation in speakers' bureaus; membership, employment, consultancies, stock ownership, or other equity interest; and expert testimony or patent-licensing arrangements), or non-financial interest (such as personal or professional relationships, affiliations, knowledge or beliefs) in the subject matter or materials discussed in this manuscript.

Author details

Romel E. Guzmán-Alvarez[1]* and José G. Márquez-Ramos[2]

1 Institute of Food Science and Technology, Science Faculty, Central University of Venezuela, Caracas, Venezuela

2 School of Nutrition and Dietetic, Medicine Faculty, University Central of Venezuelan, Caracas, Venezuela

*Address all correspondence to: romelguzman@gmail.com

References

[1] Cempaka L. Aliwarga L. Purwos S. Penia KMT. Dynamics of cocoa bean pulp degradation during cocoa bean fermentation: Effects of yeast starter culure addition. J Math Sci. 2014; 46:14-25.

[2] Salazar LG. Aislamiento y caracterización de microorganismos durante el proceso de fermentación de Theobroma caao L. de la variedad "Chuncho" obtenida en Cuzco [thesis] Perú: Universidad Cayetano Heredía; 2017

[3] Schwan R, Wheals AE. The microbiology of cocoa fermentation and its role in chocolate quality. Crit Rev Food Sci Nutr. 2004; 44: 205-221.

[4] Schwan RF, Fleet GH. Cocoa and coffee fermentations. CRC Press.Boca Raton. FL. 2014

[5] Thanh PB, Tru NV, Thi Thuy D, Thi Thoa N, Thao PV, Tham TT, Van Vo T. Bacteria in Wooden Box Fermentation of Cocoa in Daklak. Vietnam. J. Microbiol Exp. 2017; 5: 1-5.

[6] Papalexandratou Z, Kaasik K, Villagra LK, Skorstengaard A, Bouillon G, Espensen JL, Hansen LH, Jakobsen RR, Krych AB, Castro-Mejia JL, Nielsen DS. Linking cocoa varietals and microbial diversity of Nicaraguan fine cocoa quality appreciation. Int J Food Microbiol. 2019; 304:106-118

[7] Kostinek M, Koffi LB, Ottah-Atikpo M, Teniola D, Schillinger U, Holzaplel WH, Franz CMP. Diversity of predominant lactic acid bacteria associated with cocoa fermentation in Nigeria. Curr Microbiol. 2011; 56:306-314.

[8] Figueroa-Hernández C, Mota-Gutiérrez J, Ferrocino I, Hernández-Estrada ZJ, González-Ríos O, Cocolin L, Suárez-Quiroz ML. The challenges and perspectives of the selection of starter cultures for fermented cocoa beans. Int J Food Microbiol. 2019; 301: 41-50.

[9] Daniel HM, Vracken G, Takrama JF, Camu F, de Vos P. Yeast diversity of Ghanian cocoa bean heap fermentations. FEMS Yeast Research. 2009; 9: 774-783

[10] Ardhana M, Fleet G. The microbial ecology of cocoa bean fermentations in Indonesia.Int J Food Microbiol. 2003; 86:87-99.

[11] Leal GA, Gomes LH, Efraim P, De Alemida FC, Figueira A. Fermentation of cocoa (*Theobroma cacao* L.) seeds with a hybrid Kluyveromyces marxianus strain improved product quality attributes. FEMS Yeast Res. 2008; 8:788-798.

[12] Lefeber T, Janssens M, Camu N, De Vuyst L. Kinetic analysis of strains of lactic acid bacteria and acetic acid bacteria in cocoa pulp simulation medía toward development of a starter culture for cocoa bean fermentation. Appl. Environ Micobiol. 2010; 76:7708-7716.

[13] Petyaev IM, Bashmakov YK. Cocobiota: Implications for Human Health. J Nutr Metabol. 2016; ID 7906927.

[14] Guzman R, Pérez E, Raymundez M. structural and chemical changes during the fermentation of cocoa beans. In: **Miki Løvstrøm, editors.** *Theobroma cacao* Production, Cultivation and Uses. Nova Science Publishers, Inc; 2020. p. 81-104.

[15] Nielsen DS, Honholt D, Debrah T, Jespersen L. Yeast populations associated with Ghanaian cocoa fermentations analysed using denaturing gradient gel electrophoresis (DGGE). Yeast. 2005; 22: 271-284.

[16] Elkori S. Estudio de la composición química y microbiológica durante el proceso de fermentación del cacao de Chuao [Thesis]. Caracas: Universidad Central de Venezuela; 2006.

[17] Paz AE. Caracterización del proceso de fermentación de cacao (*Theobroma cacao* L.) venezolano variedad carenero [Thesis]. Caracas: Universidad Simón Bolívar; 2012.

[18] Wacher MC. Microorganismos y chocolate. Revista digital universitaria. 2011; 12: 1-9.

[19] Kouame LM, Goualie BG, Adom JN, Koua G, Ouattara HG, Dove G, Niamké SL. Diversity of microbial strains involved in cocoa fermentation from Sud. Comoe (Cote d'Ivoire) based on biochemical properties. European Scientific Journal. 2015; 11: 69-85.

[20] Koffi, AS, Yao NG, Bastide P, Bruneau D, Kadjo D. Homogenization of Cocoa Beans Fermentation to Upgrade Quality Using an Original Improved Fermenter. Int. J. Biol. Biomol. Agric. Food Biotechnol. Eng. 2017; 11: 558-563.

[21] Kurtzman CP, Fell JW, Boekhout T, Robert V. 2011. Methods for isolation, phenotypic characterization and maintenance of yeasts. The Yeasts. 2011; Vol. 1. Elsevier B.V.

[22] Jamili N, Yanti A, Susilowati PE. Diversity and the role of yeast in spontaneous cocoa vean fermentation from Southeast Sulawesi, Indonesia. Biodiversitas. 2016; 17: 90-95.

[23] Lagos JF. Identificación y caracterización de levaduras fermentadoras de cacao (*Theobroma cacao*) provenientes de centros de acopio de dos localidades del Ecuador [Thesis]. Pontificia Universidad Catolica del Ecuador; 2017.

[24] Magalhaes IM, da Cruz MG, Ferreira WD, Ribeiro DD, Freitas RS. Microbial succession and the dynamics of metabolites and sugars during the fermentation of three different cocoa (*Theobroma cacao* L.) hybrids. Food Res Int. 2013; 54: 9-17.

[25] Yang S, Pan C, Hurst GB, Dice L, Davison BH, Brown SD. Elucidation of Zymomonas mobilis physiology and stress responses by quantitative proteomics and transcriptomics. Front Microbiol. 2014; 5: 246

[26] Reis RV, Bassi AP, Silva JC, Ceccato-Antonini SR. Characteristics of Saccharomyces cerevisiae yeasts exhibiting rough colonies and pseudohyphal morphology with respect to alcoholic fermentation. Brazilian Journal of Microbiology. 2013; 44: 1121-1131

[27] Arana-Sánchez A, Segura-García LE, Kirchmayr M, Orozco-Ávila I, Lugo-Cervantes E, Gschaedler-Mathis A. Identification of predominant yeasts associated with artisan Mexican cocoa fermentation using culture- dependent and culture-independent approaches. World J Microbiol Biotechnol. 2015; 31:359-369.

[28] Penia MTA, Suryani L, Affifah M. Improvement of Cocoa Beans Fermentation by LAB Starter Addition. J Med Bioeng. 2013; 2: 274-278.

[29] Illeghems K, De Vuyst L, Papalexandratou Z, Weckx S. Phylogenetic Analysis of a Spontaneous Cocoa Bean Fermentation Metagenome RevealsNew Insights into Its Bacterial and Fungal Community Diversity. PLoS ONE. 2012; 7: e38040.

[30] Adiko EC, Ouattara HD, Dové GG, Niamké SL. Assessment of the diversity of lactic acid bacteria involved in cocoa fermentation of six main cocoa producing regions of Cote d'ivoire. Annual Research and Review Biology. 2018; 27: 1-16.

[31] Sánchez A, De Revel G. Antalick G, Herrero M, García LA, Díaz M. Influence of controlled inoculation of malolactic fermentation on the sensory properties of industrial cider. Journal of Industrial Microbiology and Biotechnology. 2015; 41: 853– 867.

[32] Agyrifo DS, Wamalva M, Otwe EO, Galyuon I, Runo S, Takrama J, Ngeranwa J. Metagenomics analysis of cocoa bean fermentation microbiome identifying species diversity and putative functional capabilities. Helyon, 2019; 5: eo2170.

[33] Papalexandratou Z, Kaasik K, Villagra LK, Skorstengaard A, Bouillon G, Espensen JL, Hansen LH, Jakobsen RR, Krych AB, Castro-Mejia JL, Nielsen DS. Linking cocoa varietals and microbial diversity of Nicaraguan fine cocoa quality appreciation. Int J Food Microbiol. 2019; 304:106-118.

[34] De Melo GV, Texeira KM, Gonzaga EM, Da Silva IC, Freitas RS. Spontaneous cocoa bean fermentation carried out in a novel-design stainlesssteel tank: Influence on the dynamics of microbial populations and physical–chemical properties. Int J Food Microbiol. 2013; 121-133.

[35] Visentin G, De Marchi M, Berry PD, McDermott A, Fenelon MA, Penasa M, McParland S. Factors associated with milk processing characteristics predicted by mid-infrared spectroscopy in a large database of dairy cows. J Dairy Sci. 2016; 100: 3293-3304.

[36] Breuer U, Harms H. Debaryomyces hansenii — an extremophilic yeast with biotechnological potential. Yeast. 2006; 23: 415-437. doi:10.1002/yea.1374

[37] Camu N, De Winter T, Verbrugghe K, Cleenwerck I, Vandamme P, Takrama JS, Vancanneyt M, De Vuyst L. Dynamics and Biodiversity of Populations of Lactic Acid Bacteria and Acetic Acid Bacteria Involved in Spontaneous Heap Fermentation of Cocoa Beans in Ghana. Applied and Environmental Microbiology. 2008; 88: 1809-1824.

[38] Batista NN, Ramos CL, Dias DR, Pinheiro ACM, Schwan RF. Dynamic behavior of Saccharomyces cerevisiae, Pichia kluyveri and Hanseniospora uvarum during spontaneous and inoculated cocoa fermentations and their effect on sensory characteristic of chocolate Food Sci Technol. 2015; 63: 221-227.

[39] Fernández Y, Balzarini T, Clapé BP, Evrard P, De Vuyst L, Daniel HM. The environmental and intrinsic yeast diversity of Cuban cocoa bean heap fermentations. Inter J Food Microbiol. 2016; 233: 34-43

[40] Miescher S, Leischtfeld F, Demarchi GC. High-throughput identification of the microbial biodiversity of cocoa bean fermentation by MALDI-TOF MS. Lett Appl Microbiol. 2016; 63: 347-355.

[41] De Vuyst L, Weckx. The cocoa bean fermentation process: from ecosystem analysis to starter culture development. J Appl Microbiol. 2016; 121: 5-17.

[42] Meersman E, Steensels J, Mathawan M, Pieter-Jan W, Saels V. Detailed analysis of the microbial population in Malaysian spontaneous cocoa pulp fermentations reveals a Core and Variable microbiota. PLoS ONE. 2013; 8: e81559.

[43] Papalexandratou Z, De Vuyst L. Assessment of the yeast species diversity of cocoa bean fermentations in different cocoa-producing regions using denaturing gradient gel electrophoresis. FEMS Yeast Research. 2011; 11: 564-574.

[44] Ravelomanana R, Guiraud JP, Vincent JC, Galzy P. The yeast flora of cocoa bean fermentation in the Ivory Coast. MIRCEN. 1985; 1: 319-326.

[45] Guehi T, Dadie A, Koffi K, Dabonne S, Ban-Koffi L, Kedjebo K, Nemlin G. 2010. Performance of different fermentation methods and the effect of their duration on the quality of raw cocoa beans. Int J Food SciTechn. 2010; 45: 2508-2514.

[46] De Melo GV, Texeira KM, Freitas RS. rDNA-based DGGE analysis and electron microscopic observation of cocoa beans to monitor microbial diversity and distribution during the fermentation process. Food Res Int. 2013; 53: 482-486.

[47] Walker G. Yeast: Physiology and biotechnology. John Wiley and Sons Ltd; 2008.

[48] Galvez SL, Loiseau G, Paredes JL, Barel M, Guiraud JP. Study on the micro flora and biochemistry of cocoafermentation in the Dominican Republic. Int J Food Microbiol. 2007; 114: 124-130.

[49] Ouattara HG, Reverchon S, Niamke SL, Nasser W. Molecular identification and pectatelyase production by Bacillusstrains involved in cocoa fermentation. Food Microbiol. 2011; 28: 1-8.

[50] Cleenwerck I, Camu N, Engelbeen K, Winter T, Vandemeuleboecke K, De Vos P, De Vuyst L. Acetobacter ghanensis sp. nov., a novel acetic acic bacterium isolated from traditional heap fermentations of Ghanaian cocoa beans. Int J Sys Evol Microbiol. 2007; 57: 1647-1652.

[51] Comisión Venezolana de Normas Industriales (COVENIN). Granos de cacao, prueba del corte. FONDONORMA. Norma 442. Caracas. 1995. http://www.sencamer.gob.ve/sencamer/normas/442-95.pdf.

[52] Comisión Venezolana de Normas Industriales (COVENIN). Norma venezolana granos de cacao. Fondonorma, Caracas. 2010. Venezuela.http://www.sencamer.gob.ve/publicos/descargas/pdf/covenin/COVENIN_50-2018.pdf

Chapter 8

Aroma Profile of Arabica Coffee Based on Ohmic Fermentation

Reta Reta, Salengke Salengke, Mursalim Mursalim, Junaedi Muhidong, Sitti Nurmiah, Ophirtus Sumule and Fitri Fitri

Abstract

Aromatic components contained in coffee are the important components. Several technologies can be used to improve the aroma quality of coffee, for example with ohmic technology. This study established a specialty coffee processing system focused on ohmic-based fermentation technology. The aim of this study was to identify the aroma compound in coffee that fermented by ohmic technology. The SPME method by gas chromatography-mass spectrometry is used in this study. The temperatures (30, 35, and 40°C) and fermentation time were set for this study (2, 6, 12, and 18 h). The results of the sensory test of sample coffee from Indonesia with specification of areas of origins Enrekang and Gowa in comparison with a sample of coffee from Japan can conclude that the panelist provided a rating profile liking the sample coffee from Japan, but the overall results of the quantitative descriptive analysis (QDA) of the second sample are similar or nonsignificantly different. Compound pyrazine identified in GC-MS is earthy odor which has a correlation with the results of sensory taste QDA on sample coffee from Japan.

Keywords: coffee, sensory, aroma, flavor, fermentation

1. Introduction

Coffee is one of the world's most widely consumed beverages. Coffee is one of the most important agricultural commodities in the world and is the most traded commodity globally after petroleum [1]. In 2012, the total value of coffee industry was estimated at US $173.40 billion [2]. The world coffee production showed a slight increase from 8.21 million metric tons in 2007 to 8.3 million metric tons in 2011 [3]. Based on data from AICE and ICO [4], Indonesia accounted for approximately 8.9% of total production in 2013.

Aromatic components contained in coffee beverages are the important components, which are affected by several factors; the sensory characteristic of coffees depends on crop varieties, growing areas, and processing methods. The processing methods that can significantly affect sensory characteristic include the methods of fermentation of the coffee fruit (from the coffee cherry to the green coffee), roasting, grinding, and brewing. Some research have been done to find the volatile compounds associated with the aromas and flavors contained in coffee [5]. In addition to environmental and agronomic factors, flavors of coffee are also strongly influenced by the way of processing, especially the process of fermentation and roasting [6–11]. Coffee handling process in general can be distinguished as wet and dry handling. Differences in the processing of coffee will produce different flavors, mainly due to their effects on chemical composition of the processed coffee [12–14]. Fermentation is instrumental in the formation of flavor-forming precursor compound, while the roasting process is known to contribute to the formation of volatile compounds and nonvolatile compounds that produce to the distinctive flavor in coffee [15].

The quality of coffee flavor can be different perceived by consumers. According to Leroy et al. [16], taste can be measured with the senses and can be influenced by the physical, chemical, agronomic, and technological factors. Aroma and flavor are undoubtedly important hedonic aspects of a good coffee [17], and thus these two aspects should be carefully considered in coffee classification.

Flavor, which is composed of taste and aroma, is very important for the quality of coffee. Describing the flavor of coffee is a very complex task, because it is influenced by various factors including on farm factors [17]. Not surprisingly, the diversity of flavors available commercially in specialty coffee is amazingly spacious such that a generic description of "coffee flavor" is impossible. A variety of studies have been performed over the last decade to investigate the basic composition of coffee flavor and nonvolatile components using advanced technology and instrumentation [18–24] and volatile compounds [25–29]. These study the chemical components of coffee.

To better understand and explore the taste of coffee, much efforts should be focused on objectively measuring sensory properties of coffee using a scientific approach. Sensory evaluation was done to characterize the sensory properties of products through the measurement of human response with minimum bias [30] and because it is the most rigorous approach to any assessment of the quality of the coffee.

Sensory evaluation (organoleptic) is a method performed by humans using human senses, namely the eyes, nose, mouth, hands, and ears. Through reviews, these five basic senses, we can assess the sensory attributes of a product such as color, form, shape, taste, and texture.

Therefore, in this study, organoleptic evaluation of QDA method is conducted by using nine types of Arabica coffee samples with different fermentation treatments and originates from coffee-producing areas in Indonesia, namely Enrekang and Gowa, and compared it to the coffee samples from Japan. The QDA method is a method in which the panel chooses the words that express the characteristics of the sample and quantifies them, gives a score to each sensory characteristic, and allows the characteristics of the sample to be visualized.

This study was conducted to identify the aroma compound in coffee, as an ingredient of the most popular beverages in the world. The main method used to identify the volatile compounds that contribute to coffee flavor is solid-phase microextraction (SPME) method by gas chromatography-mass spectrometry (GC-MS).

Solid-phase microextraction (SPME) has been the preferred tool for many coffee aroma volatile analyses in recent decades [31–42], mainly due to its sensitivity, rapidness, and solvent-free properties [43]. The extraction of samples from the headspace (HS) was also recorded as the most accurate composition of the flavors [37]. The main purpose of this study was to analyze aroma profile of Arabica after fermentation in an ohmic heater.

2. Material and methods

Coffee samples used in the test were Arabica coffee from Indonesia, from two different growing areas in the province of South Sulawesi, namely Enrekang and

Gowa. The samples tested were fermented at different conditions, namely fermentation for 2, 6, and 18 h at temperatures of 30, 35, and 40°C. As comparison, we also can tests using commercial coffee samples from Japan.

Information to using symbol sample (E, M, J = Area sample, H (time fermentation), T (Temperature fermentation)) in QDA:

A = EH2T30
B = EH2T35
C = EH12T35
D = EH18T35
E = MH2T30
F = MH2T35
G = MH6T35
H = MH18Y40
I = JH0T0

Sensory testing was performed on the sample used in the study, with three replications for each sample, from a total sample of nine.

2.1 Sample preparation techniques

2.1.1 Sample sensory test

Sample coffee in the ground is by means of ground coffee types and coffee MORNING MATE MILL CO-10. Sample ground coffee, which has been prepared in a glass with each weighing 2 g/cup to test smile, was right to test the beverage weighed 2 g/cup by cup size used which is mixed with hot water 3 oz. The test results of sensory tests proceed with testing sensory quantitative descriptive analysis.

2.1.2 Preparation sample HS-SPME

HS-SPME Coffee Powder: 4 g of powder was weighed in a septic-sealed gas vial (20 mL); the resulting headspace was sampled for 40 min at 40°C with a 350 rpm agitation rate using a PDMS/DVB SPME fiber. By sampling 5 L of a 1000 mg/L solution of n-C13 in DBP in a 20 mL headspace vial and agitating it at 3 mL for 20 min at 50°C, the internal norm was preloaded to the fiber.

2.2 GC-MS analysis

GC-MS analysis was performed using a Hew Packard 6890 N (Agilent Technologies, Inc., Santa Clara, CA) HS-SPME-GC-MS at chromatographic conditions: injector temperature: 2300°C; injection mode: splitless; carrier gas: helium (2 mL/min); fiber desorption time and reconditioning: 5 min; column: SGE SolGelwax (100 percent polyethylene glycol) 30 m × 0.25 mm; ionization mode: EI (70 eV); scan range: 35–350 amu; ion source temperature: 2000 V with liquid carbon dioxide; MS conditions: ionization mode: EI (70 eV); scan range: 35–350 amu; and ion source temperature: 2000 V.

2.3 Volatile component identification

Aroma compounds were classified by comparing their measured linear retention indices and mass spectra to those of genuine samples or, provisionally, those obtained from home-made or commercial libraries (Wiley 7 N and Nist 05 ver 2.0 Mass spectral data) or published in the literature.

2.4 Quantitative descriptive sensory analysis

The approach of quantitative descriptive analysis (QDA) is based on the concept of the ability of panelists to verbalize impressions of the commodity in a reliable manner. The process embodies a systematic screening and training technique, the creation and use of a sensory language, and the labeling of items in a repeat test in order to achieve a complete, quantitative definition. We followed the method of stone and side, a panel of 10 assessors (5 males and 5 females, age range 20–50 years). Initially, 90 descriptive attributes to coffee beverage were Aroma (attributes, and to test beverage, get the six attributes after getting consensus among 10 assessors on the definition of terms as shown in **Figure 1**). The rating scale method shown below was evaluated by using a two-way analysis of variance (ANOVA).

2.5 Sensory test with a rating scale method

With a line scale, the members of the panel rated the strength of the coffee aroma by labeling the horizontal line. The marked point corresponded to the influence of the stimulus perceived. The length of the line was 14 cm with anchors on both ends. The left end of the scale corresponded to "none," while the right end corresponded to such descriptive terms. The panel consisted of 5 males and 5 females. Their ages ranged from 20 to 50 years with a mean of 24 years.

2.6 Statistical analysis

Analysis of variance (ANOVA) was applied for the determination of the main effects of the investigated independent factors (flavor and percentage) on the sensory attributes as they were rated during the QDA procedure. Duncan's multiple range test was used to separated means of sensory attributes when significant differences ($p < 0.05$) were observed.

A Shimadzu GC-MS solution of 2.5 SU1 and an Agilent ChemStation D.02.00.275 were used to collect data. To visualize sample groups and compare information provided by each sampling, Principal component analysis (PCA) was used. In normalized ISTD data, PCA based on Pearson correlation coefficient was

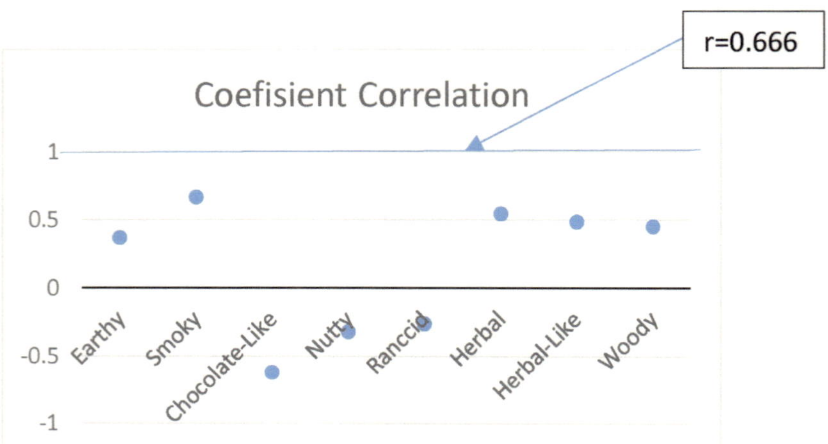

Figure 1.
Graphical representation of coefficient correlation characteristic attribute.

performed. XLSTAT (version 2015.5.01.23164) copyright Addinsoft 1995–2015 was used for statistical analysis, including one-way ANOVA and PCA. Non-polarity.

3. Results and Discussion

3.1 Quantitative descriptive analysis of various kinds of coffee brewed

A panel of 10 assessors was selected to evaluate 8 aroma attributes and 5 taste attributes of coffee beans. The aroma attributes tested are earthy, smoky, chocolate, caramel, nutty, rancid, herbal and woody. While the taste attributes tested are acidity, bitter, sweet, salt, and sour.

The strength of the correlation between the characteristic expression term and the taste (preference) that had a significant difference was calculated using the correlation function in Excel (**Table 1**).

From **Table 1**, when n = 9, from the correlation coefficient test table, it is judged that there is a significant positive correlation when r = 0.666 or more at 5% risk rate. So, it is shown from **Table 1** data of coefficient correlation result that all attributes have no correlation with the liking, because all the number of r is 0.05 from the table (n = 9 and r total sample = 0.666) which is bigger than the number of the correlation result (see **Figure 1**).

Figures 1 and 2 show the smell of sample coffee from Indonesia (sample difference treatment, fermentation) and Japan, so is the result is gaining attribute earthy, smoky, chocolate-like, caramel-like, nutty, herbal-like, and woody, like all sample are similar because the number from the result analysis, statistical (ANOVA), is low when compared to the number from q table, when total sample is 9 (q9 72 = 4.7937), and except the attribute rancid, only for sample A, C, G, I, and F is similar, then sample B, E, D, and H is not similar because the number from result ANOVA is bigger than number from q table.

Figure 3 shown is the taste of coffee sample from Indonesia (sample difference treatment, fermentation) and Japan, so is the result attribute bitterness, sweetness, saltiness, like all samples are similar because the number from result analysis, statistical (ANOVA), is low compared to the number from q table, when total sample is 9 (q9 72 = 4.7937), and except the attribute acidity for the sample D, I, and F are not similar because the number is bigger than number from table q, and then attribute sourness for the sample I, D, and F is same as the attribute acidity is not similar.

Attributes	Coefficient correlation
Earthy	0.37155974
Smoky	0.664932567
Chocolate-like	−0.617978767
Nutty	−0.316895337
Rancid	−0.263026567
Herbal	0.544261264
Herbal-like	0.486931132
Woody	0.451859838

Table 1.
Correlation coefficient of each characteristic expression term.

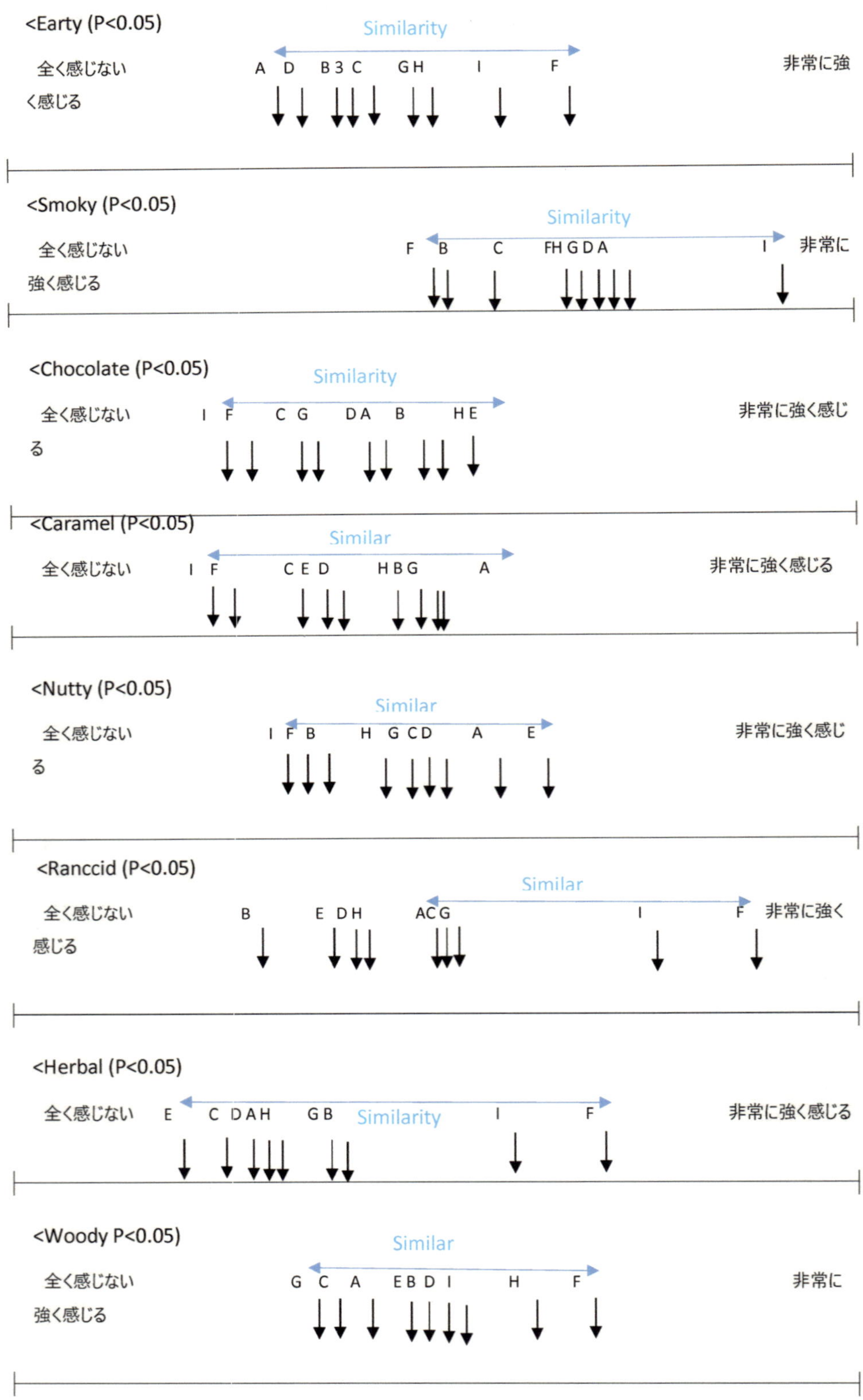

Figure 2.
Sensory analysis of coffee based on earthy, smokey, chocolate-like, caramel like, nutty, rancid, herbal-like, and woody-like odor flavor attributes.

Figure 4(A) shows the results of sensory taste with QDA described in the sample A with smokey attribute panelist, giving the highest value in the sample compared with a sample of coffee from Japan than Indonesia. Attribute woody and liking panelist provide similar valuation between the two samples of the coffee, but the attribute rancid panelist provides high value, attribute nutty, caramel, and chocolate panelist, giving nonsimilarity to the second sample.

Based on the results, it can be seen that coffee with sample code A (EH2T30) has the highest value on the sour taste attribute, while coffee with code I (Japan) has the highest value on the sweet taste attribute. This shows that coffee A and coffee I have no similarities, and in general the panelists prefer coffee with sample code I.

Figure 4(A) and **(B)** can be explained that the sensory panelists' smell gives the highest value apart from all attributes except chocolate, caramel, and nutty, but at taste, sensory panelists provide the highest value of the sample A except attribute liking.

Based on the sensory analysis of the aroma of coffee beans, it shows that the woody and herbal aroma attributes in coffee sample I and B have similarities, but the smoke and rancid attributes in sample B have a higher value than coffee sample I. The taste attributes in both coffee samples show that sample B coffee has higher acidity and sour attributes than sample I coffee. There are similarities in the value of the sweet attribute in coffee I and coffee B.

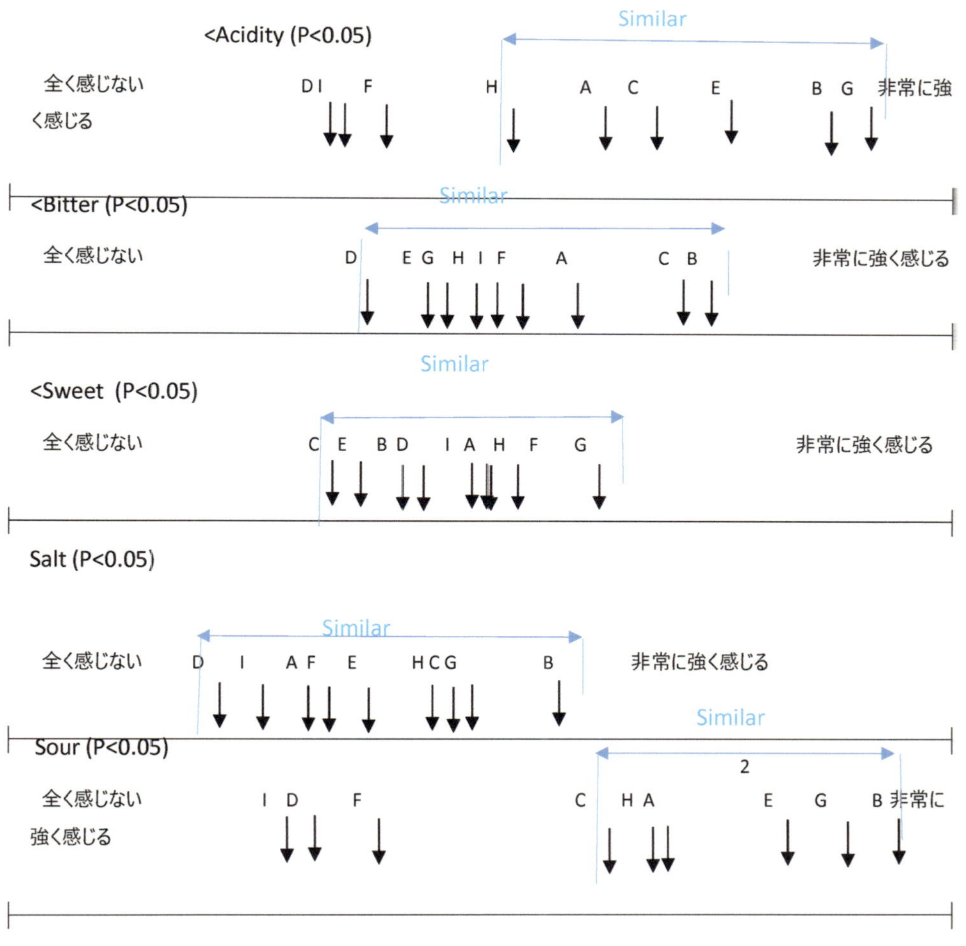

Figure 3.
Sensory analysis of coffee based on acidity, bitterness, sweetness, saltiness, sourness, and liking flavor attributes.

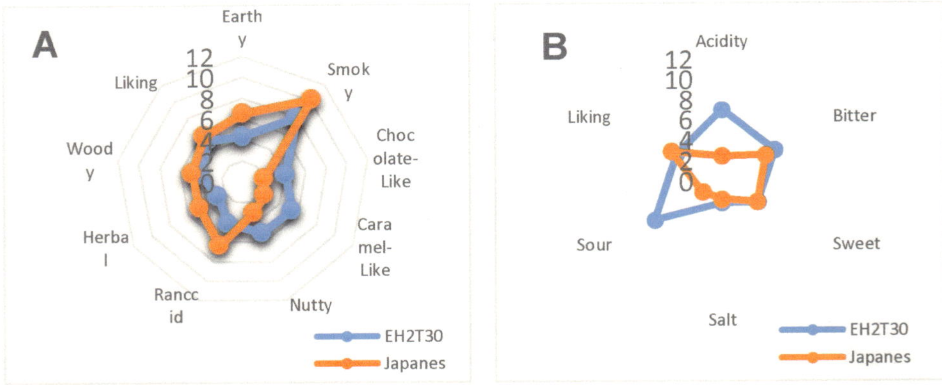

Figure 4.
(A) Result of sensory aroma sample a (EH2T30) with coffee from Japan (I) (JHoTo), and (B) result of sensory tastes sample a (EH2T30) with coffee from Japan (I).

Figure 5(A) showed the sample coffee B (EH18T35) than I result of QDA and sensory test is nonsimilar for the attribute smoky, rancid, and herbal, because the number of attributes is bigger than sample B (EH18T35), and the attribute that is similar is the nutty, caramel-like, and chocolate-like. And **Figure 5(B)** showed the sample B (EH18T35) and I being nonsimilar for attributes acidity, sourness, and saltiness.

Figure 6(A) showed the sample coffee B (EH18T30) than I result analysis for attribute sweetness and liking is similar, and all attributes for a sample coffee B (EH18T30) number is getting bigger than the sample I for the sensory tastes.

Figure 7(A) showed that the results of QDA and sensory aroma of sample coffee D (EH2T35) and I such as smokey and rancid is nonsimilar. High scores aroma smoke and rancid were also found in sample with I (Japan coffee).

The lowest aroma scores were found in sample D (EH2T35) as attributes being earthy, herbal, and rancid. **Figure 7(B)** showed the sensory tastes of coffee sample D (EH2T35) and I were found to be similar except the bitterness being lowest taste flavor.

Figure 8(A) from QDA of each attribute showed nonsimilar (significant difference) in smokey, rancid, herbal, nutty, and caramel-like when smelled directly. And **Figure 8** showed also non similar in acidity, sourness when tastes directly. Sample E (MH2T35) showed the lowest value in earthy, rancid, smoky, herbal and liking. No significant difference was noted by 10 panelists.

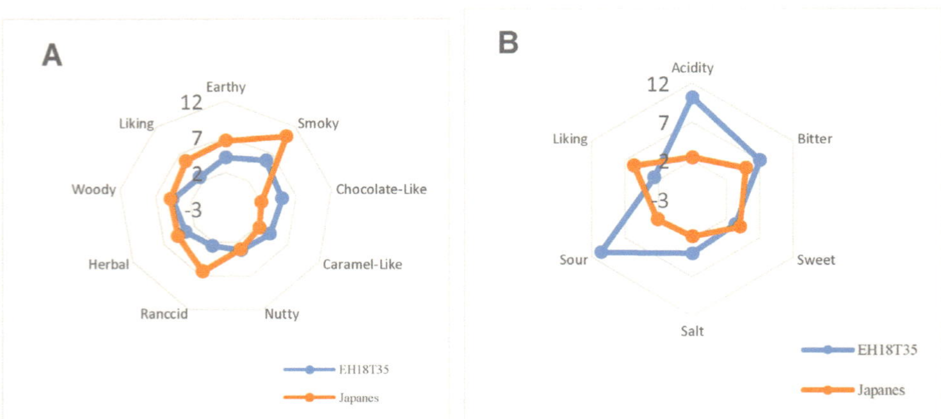

Figure 5.
(A) Result of sensory aroma sample B (EH18T35) with coffee from Japan (I), and(B) result of sensory tastes sample B (EH18T35) with coffee from Japan (I).

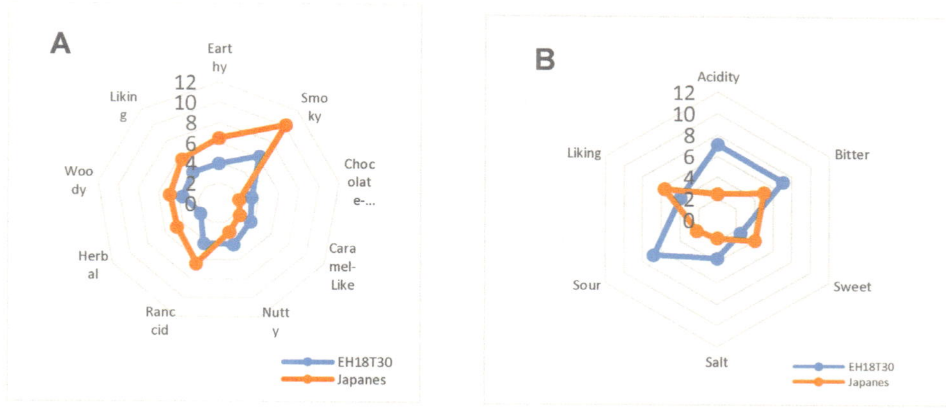

Figure 6.
(A) Result of sensory aroma sample C (EH18T30) with coffee from Japan (JHoTo), and (B) result of sensory tastes sample C (EH18T30) with coffee from Japan (I).

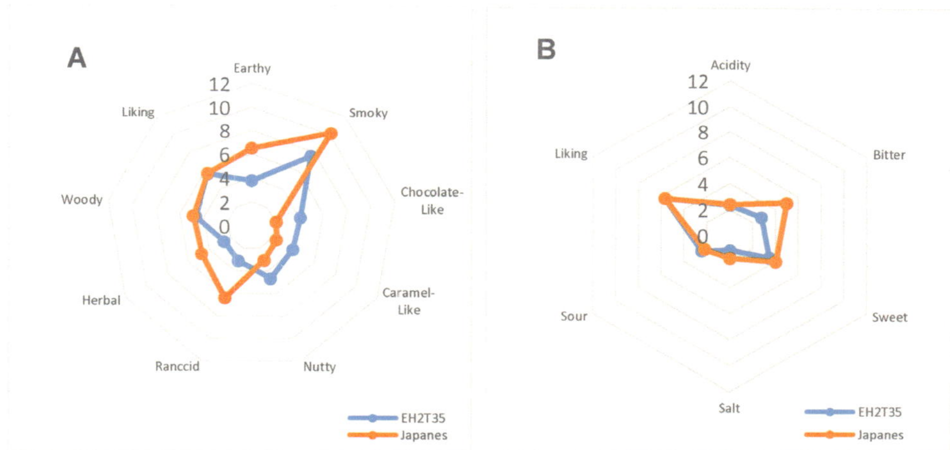

Figure 7.
(A) Result of sensory aroma sample D (EH2T35) with coffee from Japan (I), and (B) result of sensory tastes sample D (EH2T35) with coffee from Japan (I).

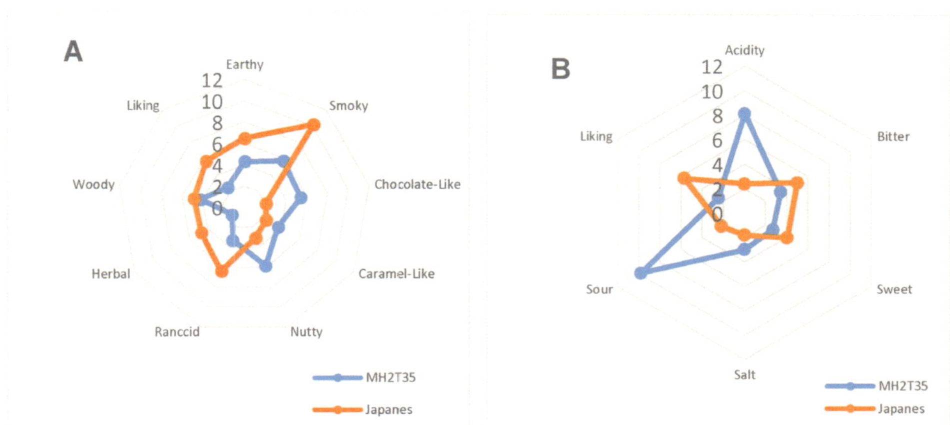

Figure 8.
(A) Result of sensory aroma sample E (MH2T35) with coffee from Japan (I), and (B) result of sensory tastes sample E (MH2T35) with coffee from Japan (I).

When comparing to type sensory analysis of coffee beverage, sample coffee E (MH2T35) has the lowest value than sample I (Japan), and with the other sensory tastes, the sample I (Japan) has higher preference and a higher liking attribute.

Figure 9(A), shows that all aroma attributes except the smoke attribute in sample F have similarities with the aroma attribute of sample I. The smoke attribute in sample I has a higher value than coffee sample F. **Figure 9(B)** showed all attributes of samples F and I are similar. When comparing the two types of sample coffees' smell and tastes, sample coffee F has higher value than sample I.

Figure 10(A), from QDA of each attribute, showed a nonsimilarity (significant difference) in smokey, rancid, herbal, nutty, and caramel-like when smelled directly. And **Figure 11** showed also nonsimilarity in acidity and sourness when tasted directly. Sample G showed the lowest value in earthy, rancid, smoky, herbal, and liking. No significant difference was noted by 10 panelists.

When comparing to type sensory analysis of coffee smell, sample coffee G has the lowest value than sample I, and with the other sensory tastes, sample G has

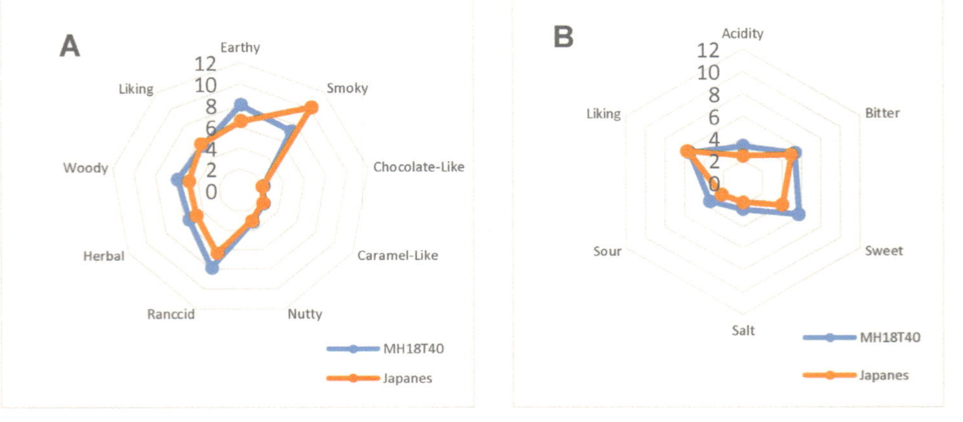

Figure 9.
(A) Result of sensory aroma sample F (MH18T40) with coffee from Japan (I), and (B) result of sensory tastes sample F (MH18T35) with coffee from Japan (I).

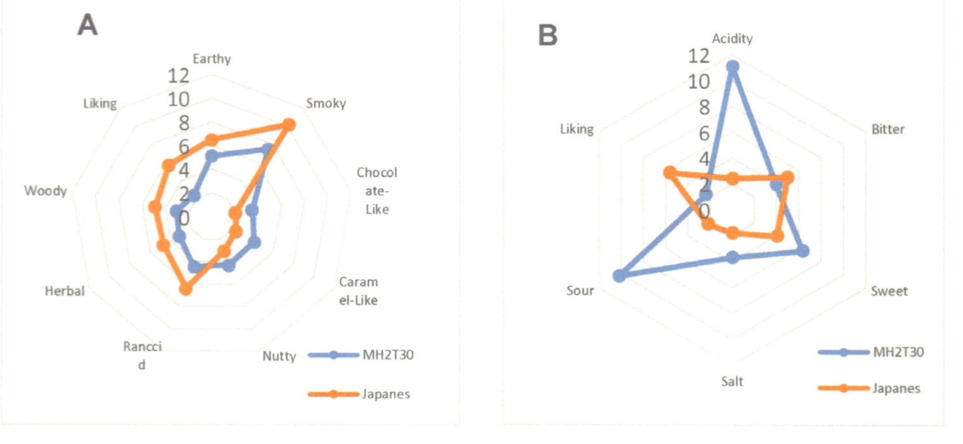

Figure 10.
(A) Result of sensory aroma sample G (MH2T30) with coffee from Japan (I), and (B) result of sensory tastes sample G (MH2T30) with coffee from Japan (I).

higher value preference for the attributes acidity and sourness than the sample I which has very lowest value due to a higher liking attribute.

Figure 11(A), from the results of QDA of each attribute, showed a nonsimilarity (significant difference) in smokey, rancid, and chocolate-like smell directly. Attribute smoke and rancid from the sample I showed higher value than sample H. **Figure 11(B)** also shows the non-similarity in the acidity, saltiness, and sourness attributes between sample H and sample I. These three attributes have higher values in sample H compared to sample I.

Based on **Figure 12**, it can be explained that the earthy aroma of coffee is the main attribute; the greatest compound that provides the earthy aroma is pyrazine; from nine samples tested, the largest sample from Japan coffee that has aroma earthy is identified. Furthermore, sample G (MH2T30) is a sample of the unidentified compound pyrazine, but the results of sensory taste panelists can identify the smell.

Based on **Figure 13**, it can be seen that the 2,2-Furanmethanol compound has the highest peak area of GC-MS result for the smoky attribute aroma. Based on the results of sensory taste, the panelist also found that smoky attribute is similar for all sample coffees, and gives no significant sample.

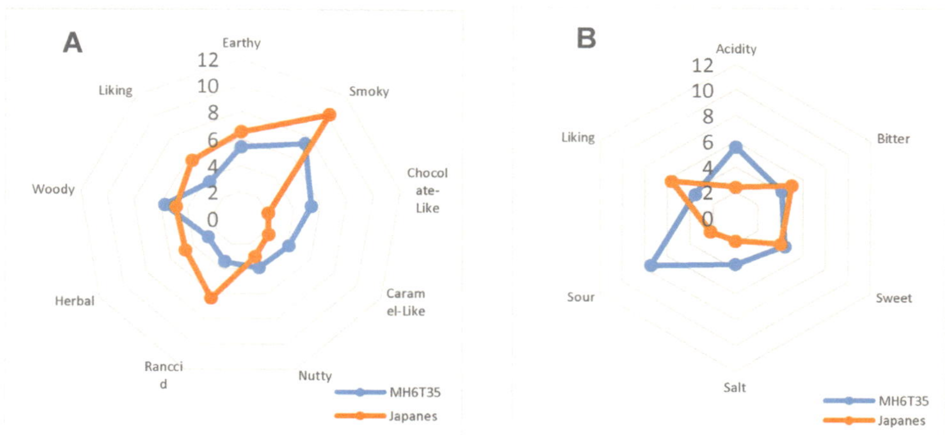

Figure 11.
(A) Result of sensory aroma sample H(MH6T35) with coffee from Japan (I), (B) Result of sensory taste sample H(MH6T35) with coffee from Japan (I).

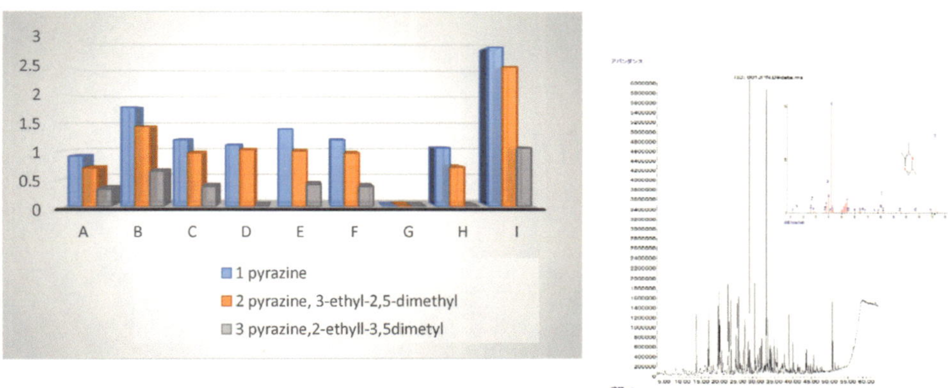

Figure 12.
Compound volatile aroma Arabica coffee by GC-MS compared with sensory taste attribute earthy.

Based on **Figure 14**, volatile compound identified from chocolate's aroma has five largest compounds, and the largest is compound pyrazine, 2,6-dimethyl in EH2T35 sample. Pyrazine-1,4-diazine compound is a compound which is identified

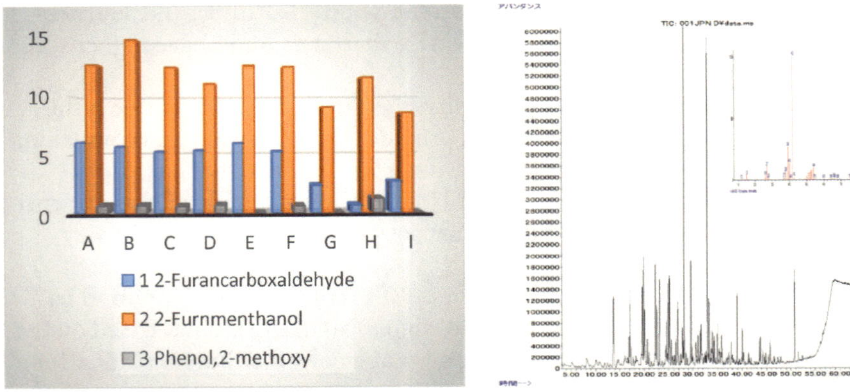

Figure 13.
Compound volatile aroma Arabica coffee bye GC-MS compare with sensory taste attribute smoky.

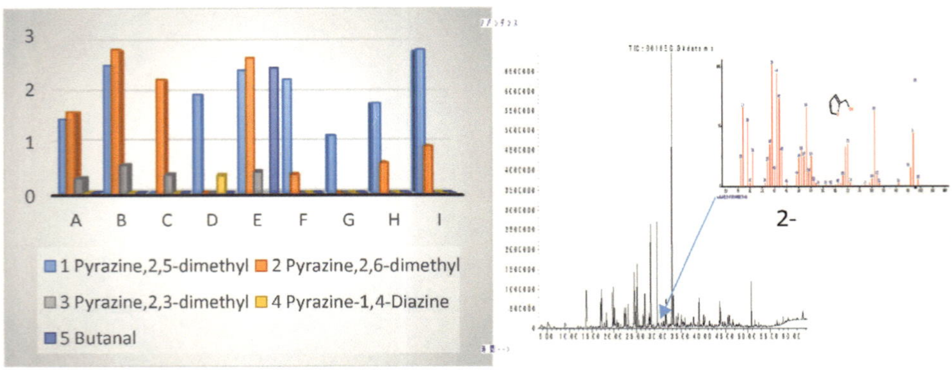

Figure 14.
Compound volatile aroma Arabica coffee by GC-MS compared with sensory taste attribute chocolate.

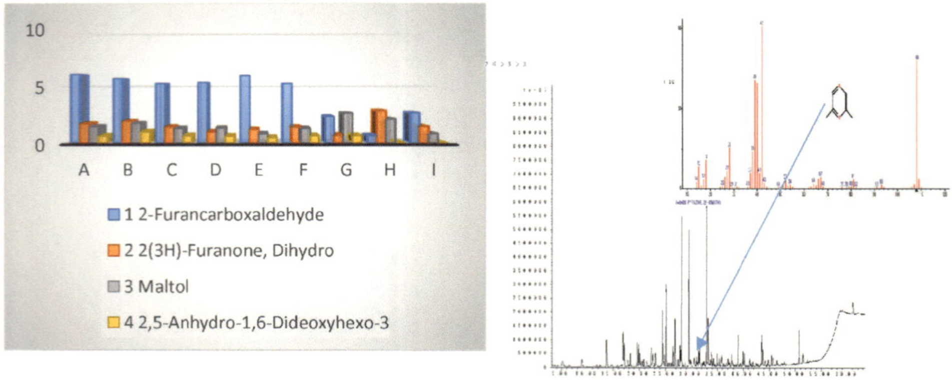

Figure 15.
Compound volatile aroma Arabica coffee by GC-MS compared with sensory taste attribute caramel.

only in the sample EH18T35, while the other sample was not identified. Sample G (MH6T35) is a compound sample that can identify only one type of its volatile compound which is pyrazine, 2,5-dimethyl.

Based on **Figure 15**, it can be explained that the volatile compound identified by GC-MS related to the results of sensory taste attributes aroma are 2-furancarboxaldehyde, 2(3H)-Furanone, Maltol, and 2,5-anhydro-1,6-dideoxyhexo. Volatile compound identified from the largest peak area is 2-Furancarboxaldehyde and is identified in sample A (EH2T30). Maltol compound is a compound which is identified in all kinds of samples of coffee with a peak area that is not too wide and not too small.

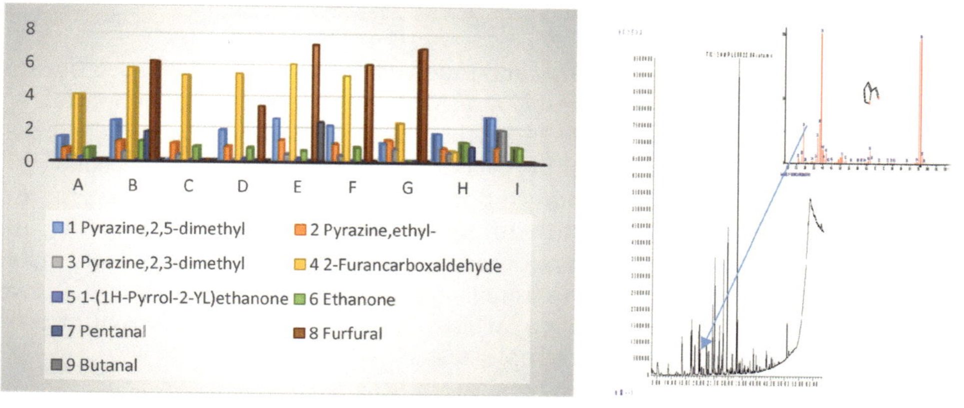

Figure 16.
Compound volatile aroma Arabica coffee by GC-MS compared with sensory taste attribute nutty.

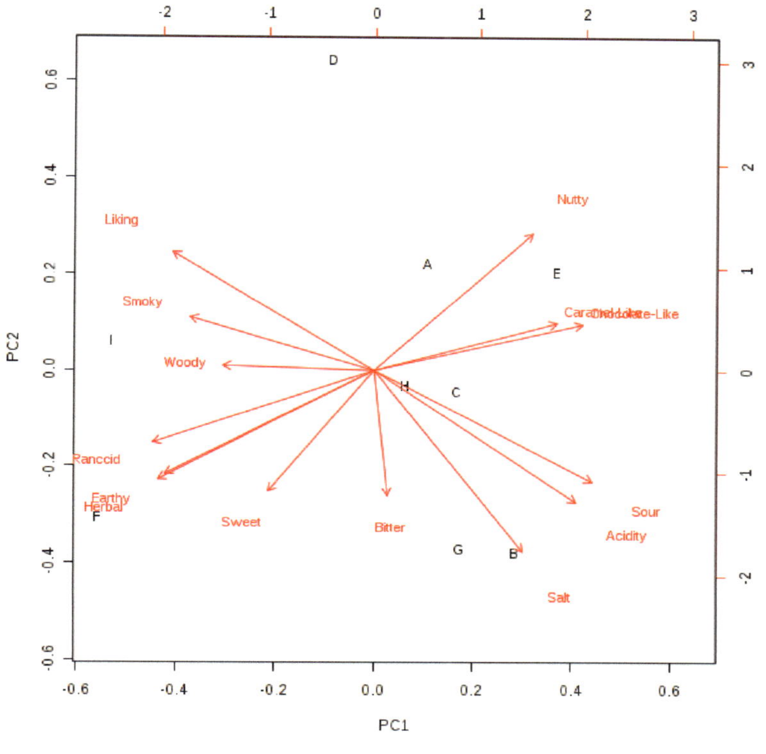

Figure 17.
Results of PCA attribute.

Attribute nutty coffee aroma in the studied sample identified by GC-MS can be seen in **Figure 16**. There are nine volatile compounds identified. The largest peak area of volatile compound identified is furfural; this compound is identified in samples F (MH2T35) with the area of the peak being 7.1846.

Based on the results of the sensory analysis in **Figure 17**, it can be seen that the panelists rated "like and have a correlation" on the smoky, woody, and rancid attributes. For the acidity attribute, the panelists assessed "dislike but have a correlation". In addition, the panelists assessed "dislike and no correlation" on the nutty, caramel-like, sweet, earthy, and herbal attributes.

4. Conclusions

From the results of sensory test, sample coffee from Indonesia with specification of areas of origins Enrekang (E) and Gowa (M) in comparison with a sample of coffee from japan (J), it can be concluded that the panelist provided a rating profile liking the sample coffee Japan, but the overall results of the QDA on the second sample are similar or nonsignificantly different. This means that the aroma profile as a whole does not provide too much difference. Pyrazine compound identified in GC-MS has an earthy odor, which has a correlation with the results of sensory taste based on the QDA on sample coffee from Japan. Reviewed nutty odor identified five compounds on GC-MS, and there is a correlation with the results of sensory taste of QDA which is similar to all coffee samples tested.

Author details

Reta Reta[1*], Salengke Salengke[2], Mursalim Mursalim[2], Junaedi Muhidong[2], Sitti Nurmiah[1], Ophirtus Sumule[3] and Fitri Fitri[4]

1 Department of Agroindustry, Pangkep State Polytechnic of Agriculture, South Sulawesi, Indonesia

2 Department of Agriculture Engineering and Food Technology, Hasanuddin University, Makassar, Indonesia

3 Department of Fisheries and Marine Science, Hasanuddin University, Indonesia

4 Agricultural Science Study Program, Graduate School, Hasanuddin University, Indonesia

*Address all correspondence to: retariskinapolitani@gmail.com

References

[1] Lee LW, Cheong MW, Curran P, Yu B, Liu SQ. Coffee fermentation and flavor—An intricate and delicate relationship. Food Chmistry. 2015;**185**:182-191

[2] International Coffee Organization. World Coffee Trade (1963-2013): A Review of the Markets, Challenges and Opportunities Facing the Sector. Breakdown of Exports of Green Arabica and Green Robusta for Countries Exporting Significant Volumes of Both Types of Coffee [Internet]. London, United Kingdom: International Coffee Organization; 2014. Retrieved from <http://dev.ico.org/prices/m1a.htm> [Accessed: March 24, 2015]

[3] FAO. Worldwide Production of Coffee Beans. 2013. Available from: http://faostat3.fao.org/faostat-gateway/go/to/home/E [Accessed: March 20, 2016]

[4] International Coffee Organization. World Coffee Trade (1963-2013): A Review of the Markets, Challenges and Opportunities Facing the Sector. Breakdown of Exports of Green Arabica and Green Robusta for Countries Exporting Significant Volumes of Both Types of Coffee. London, United Kingdom: International Coffee Organization; 2014. Retrieved from: http://dev.ico.org/prices/m1a.htm [Accessed: March 24, 2015].

[5] Czerny M, Mayer F, Grosch W. Sensory study on the character impact odorants of roasted Arabica coffee. Journal of Agricultural and Food Chemistry. 1999;**47**:695-699. DOI: 10.1021/jf980759i

[6] Avallone S, Brillouet JM, Guyot B, Olguin E, Gulraud JP. Involvement of pectolytic mikroorganisme in coffee fermentation. International Journal of Food Science and Technology. 2002;**37**:191-199

[7] Jackels SC, Jackels CF. Characterization of the coffee mucilage fermentation process using chemical indicators: A field study in Nicaragua. Journal of Food Science. 2005;**70**:321-325

[8] Pimenta TV, Pereira RGF, Correa JLG, Silva JR. Roasting processing of dry coffee cherry: Influence of grain shape and temperature on physical chemical and sensorial grain properties. Boletim do Centro de Pesquisa de Processamento de Alimentos. 2009;*27*(1):97-106

[9] Budryn G, Nebesny E, Kula J, Majda T, Kryslak W. HS-SPMJME/GC/MS profiles of convectively and microwave roasted Ivory Coast robusta coffee brews. Czech Journal of Food Sciences. 2011;**29**(2):151-160

[10] Sompom C, Kamtuo A, Theerakulpisur P, Siriamompun S. Effects of roasting degree on radical scavenging activity, phenolics and volatile compounds of arabica coffee beans. International Journal of Food Science and Technology. 2011;**46**: 2287-2296. DOI: 10.1111/j.1365-2621.2011.02748.x

[11] Wang N. Physicochemical changes of coffee beans during roasting [thesis]. Ontario, Canada: University of Guelph; 2012. Available from: http://hdl.handle.net/10214/3584

[12] Farah A, Monteiro MC, Calado F, Franca AS, Trugo LC. Correlation between cup quality and chemical attributes of Brazilian coffee. Food Chemistry. 2006;**98**:373-380. DOI: 10.1016/j.foodchem.2005.07.032

[13] Bytof G, Knopp SE, Schieberle P, Teutsch L, Selmar D. Influence of processing on the generation of gamma-aminobutyric acid in green coffee beans. European Food Research and

Technology. 2005;**220**(3-4):245-250. DOI: 10.1007/s00217-004-1033-z

[14] Taba J. Coffee Taste Analysis of an Espresso Coffee Using Nuclear Magnetic Spectroscopy [Thesis]. Eindhoven, Holland: Central Ostrobothnia University of Applied Sciences; 2012 Available from: http://urn.fi/URN:NBN:fi:amk-201204275380

[15] Jansen SO. Chemistry of coffee. In: Mender L and Liu HW, Editors. Comprehensive Natural Products II, Chemistry and Biology. United Kingdom: Elsevier Ltd.; 2014. pp. 1085-1113 Available from: https://www.elsevier.com/books/comprehensive-natural-products-ii/mander/978-0-08-045381-1

[16] Leroy T, Ribeyre F, Bertrad B, Charmetant P, Dufour M, Montagnon C. Genetics of coffee quality. Brazilian Journal of Plant Physiology. 2006;**18**:229-242. DOI: 10.1590/S1677-04202006000100016

[17] Sunarharum WB, Willianms DJ, Smyth HE. Complexity of coffee flavor: A compositional and sensory perspective. Food Research International. 2014;**62**:315-325. DOI: 10.1016/j.foodres.2014.02.030

[18] Ginz M, Balzer HH, Bradbury AGW, Maier HG. Formation of aliphatic acids by carbohydrate degradation during roasting of coffee. European Food Research and Technology. 2000;**211**:404-410 Available from: https://link.springer.com/article/10.1007/s002170000215

[19] Ginz M, Engelhardt U. Identification of new diketopiperazines in roasted coffee. European Food Research and Technology. 2001;**213**:8-11 Available from: https://link.springer.com/article/10.1007/s002170100322

[20] Farah A, De Paulis T, Trugo LC, Martin PR. Effect of roasting on the formation of chlorogenic acid lactones in coffee. Journal of Agricultural and Food Chemistry. 2005;**53**:1505-1513. DOI: 10.1021/jf048701t

[21] Farah A, Donangelo CM. Phenolic compounds in coffee. Brazilian Journal of Plant Physiology. 2006;**18**:23-36. DOI: 10.1590/S1677-04202006000100003

[22] Oestreich-Janzen S. Chemistry of coffee. In: Lew M, Hung-Wen L, editors. Comprehensive Natural Products II. Oxford: Elsevier; 2010 https://www.elsevier.com/books/comprehensive-natural-products-ii/mander/978-0-08-045381-1

[23] Liu Y, Kitts DD. Confirmation that the maillard reaction is the principle contributor to the antioxidant capacity of coffee brews. Food Research International. 2011;**44**:2418-2424. DOI: 10.1016/j.foodres.2010.12.037

[24] Bartel C, Mesias M, Morales FJ. Investigation on the extractability of melanoidins in portioned espresso coffee. Food Research International. 2015;**67**:356-365. DOI: 10.1016/j.foodres.2014.11.053

[25] Andueza S, De Pena MP, Cid C. Chemical and sensorial characteristics of espresso coffee as affected by grinding and torrefacto roast. Journal of Agricultural and Food Chemistry. 2003;**51**:7034-7039. DOI: 10.1021/jf034628f

[26] Baltes W, Knoch E. Model reactions on roast aroma formation. XIII. The formation of some uncommon N-heterocyclic compounds and furans after roasting of tryptophan with reducing sugars and sugar degradation products. Food Chemistry. 1993;**46**:343-349. DOI: 10.1016/0308-8146(93)90002-W

[27] Grosch W. Specificity of the human nose in perceiving food odorants. Paper presented at the Ninth Weurman

Flavour Research Sympodium Garchingl Deutsche forschungsanstalt fur Lebensmittelchemie. 2000

[28] Kumazawa K, Masuda H. Identification of odor-active 3-mercapto-3- methylbutyl acetate in volatile fraction of roasted coffee brew isolated by steam 159 distillation under reduced pressure. Journal of Agricultural and Food Chemistry. 2003;**51**:3079-3082. DOI: 10.1021/jf021190v

[29] Mestdagh F, Davidek T, Chaumonteuil M, Folmer B, Blank I. The kinetics of coffee aroma extraction. Food Research International. 2014;**63C**:271-274. DOI: 10.1016/j.foodres.2014.03.011

[30] Lawless HT, Heymann H. Physiological and psychological foundations of sensory function In. In: Lawless HT, Heymann H, editors. Sensory Evaluation of Food: Principles and Practices. Food Science Text Series. New York: Springer; 2010. ISBN 978-1-4615-7843-7

[31] Akiyama M, Murakami K, Ikeda M, Iwatsuki K, Wada A, Tokuno K, et al. Analysis of the headspace volatiles of freshly brewed Arabica coffee using solid-phase microextraction. Journal of Food Science. 2007;**72**:388-396. DOI: 10.1111/j.1750-3841.2007.00447.x

[32] Akiyama M, Murakami K, Ohtani N, Iwatsuki K, Sotoyama K, Wada A, et al. Analysis of volatile compounds releases during the grinding of roasted coffee beans using solid-phase microextraction. Journal of Agricultural and Food Chemistry. 2003;**51**:1961-1969. DOI: 10.1021/jf020724p

[33] Bicchi C, Drigo S, Rubiolo P. Influence of fibre coating in headspace solid-phase microextraction–gas chromatographic analysis of aromatic and medicinal plants. Journal of Chromatography A. 2000;**892**:469-485. DOI: 10.1016/S0021-9673(00)00231-4

[34] Bicchi C, Iori C, Rubiolo P, Sandra P. Headspace sorptive extraction (HSSE), stir bar sorptive extraction (SBSE), and solid phase microextraction (SPME) applied to the analysis of roasted Arabica coffee and coffee brew. Journal of Agricultural and Food Chemistry. 2002;**50**(3):449-459. DOI: 10.1021/jf010877x

[35] Bicchi C, Riosi MR, Cagliero C, Cordero C, Liberto E, Rubiolo P, et al. Quantitative analysis of volatiles from solid matrices of vegetable origin by high concentration capacity headspace techniques: Determination of furan in roasted coffee. Journal of Chromatography A. 2011;**1218**:753-762. DOI: 10.1016/j.chroma.2010.12.002

[36] Bicchi CP, Panero OM, Pellegrino GM, Vanni AC. Characterization of roasted coffee and coffee beverages by solid phase microextraction–gas chromatography and principal component analysis. Journal of Agricultural and Food Chemistry. 1997;**45**:4680-4686. DOI: 10.1021/jf9704827

[37] Caprioli G, Cortese M, Cristalli G, Maggi F, Odello L, Ricciutelli M, et al. Optimization of espresso machine parameters through the analysis of coffee odorants by HS-SPME–GC/MS. Food Chemistry. 2012;**135**:1127-1133. DOI: 10.1016/j.foodchem.2012.06.024

[38] Costa Freitas AM, Parreira C, Vilas-Boas L. The use of an electronic aroma-sensing device to assess coffee differentiation—Comparison with SPME gas chromatography–mass spectrometry aroma patterns. Journal of Food Composition and Analysis. 2001;**14**:513-522. DOI: 10.1006/jfca.2001.0987

[39] Jelen HH, Majcher M, Dziadasand Dziadas M. Microextraction techniques

in the analysis of food flavorflavor compounds: A review. Analytica Chimica Acta. 2012;**738**:13-26. DOI: 10.1016/j.aca.2012.06.006

[40] Ribeiro FC, Borém FM, Giomo GS, Lima RR, Malta MR, Figueiredo LP. Storage of green coffee in hermetic packaging injected with CO2. Journal of Stored Products Research. 2011;**47**:341-348. DOI: 10.1016/j.jspr.2011.05.007

[41] Ribeiro JS, Teófilo RF, Augusto F, Ferreira MMC. Simultaneous optimization of the microextraction of coffee volatiles using response surface methodology and principal component analysis. Chemometrics and Intelligent Laboratory Systems. 2010;**102**:45-52. DOI: 10.1016/j.chemolab.2010.03.005

[42] Ryan D, Shellie R, Tranchida P, Casilli A, Mondello L, Marriott P. Analysis of roasted coffee bean volatiles by using comprehensive two 165 dimensional gas chromatography–time-of-flight mass spectrometry. Journal of Chromatography A. 2004;**1054**:57-65 https://pubmed.ncbi.nlm.nih.gov/15553131/

[43] Pawliszyn J. Solid Phase Microextraction, Theory and Practice. New York: Wiley-VCH; 1997. ISBN: 978-0-471-19034-9

Chapter 9

From Food Waste to Volatile Fatty Acids towards a Circular Economy

Mónica Carvalheira and Anouk F. Duque

Abstract

The food industrial sector generates large amounts of waste, which are often used for animal feed, for agriculture or landfilled. However, these wastes have a very reach composition in carbon and other compounds, which make them very attractive for valorization through biotechnological processes. Added value compounds, such as volatile fatty acids (VFAs), can be produced by anaerobic fermentation using pure cultures or mixed microbial cultures and food waste as carbon source. Research on valuable applications for VFAs, such as polyhydroxyalkanoates, bioenergy or biological nutrient removal, towards a circular economy is emerging. This enhances the sustainability and the economic value of food waste. This chapter reviews the various types of food waste used for VFAs production using mixed microbial cultures, the anaerobic processes, involved and the main applications for the produced VFAs. The main parameters affecting VFAs production are also discussed.

Keywords: acidogenic fermentation, volatile fatty acids, food waste, mixed microbial cultures, process parameters, applications

1. Introduction

The increase of industrialization and world population is leading to a huge generation of organic wastes, causing serious environmental problems if disposed without an adequate treatment [1, 2]. The conventional waste treatment is mainly focused on environmental regulations, neglecting the resource recovery from wastes streams, which is one of the environmental sustainability goals [2, 3]. The resource recovery allows the waste treatment and, simultaneously, the generation of added-value products, following the circular economy strategy. The conversion of food waste (FW) into valuable products, that can be used in daily activities, have been gaining more attention due to their potential and market opportunities [4].

One of the most common technology for waste treatment is the anaerobic digestion process (composed by four stages: hydrolysis, acidogenesis, acetogenesis and methanogenesis), where the organic matter is converted into valuable resources, as methane or volatile fatty acids (VFAs) (**Figure 1**). Although biogas is generally the final product of anaerobic digestion process, the production of VFAs from FW has gained a great attention due to their high market value, as well as due to their storage and transportation be easier and safer [5]. Furthermore, the production of VFA from FW allow the replacement of the traditional production from non-renewable petrochemical sources, contributing to the circular economy and environmental sustainability [3].

VFAs are linear short-chain fatty acids comprising two (acetic acid) to six (caproic acid) carbon atoms which can be distilled at atmospheric pressure [2].

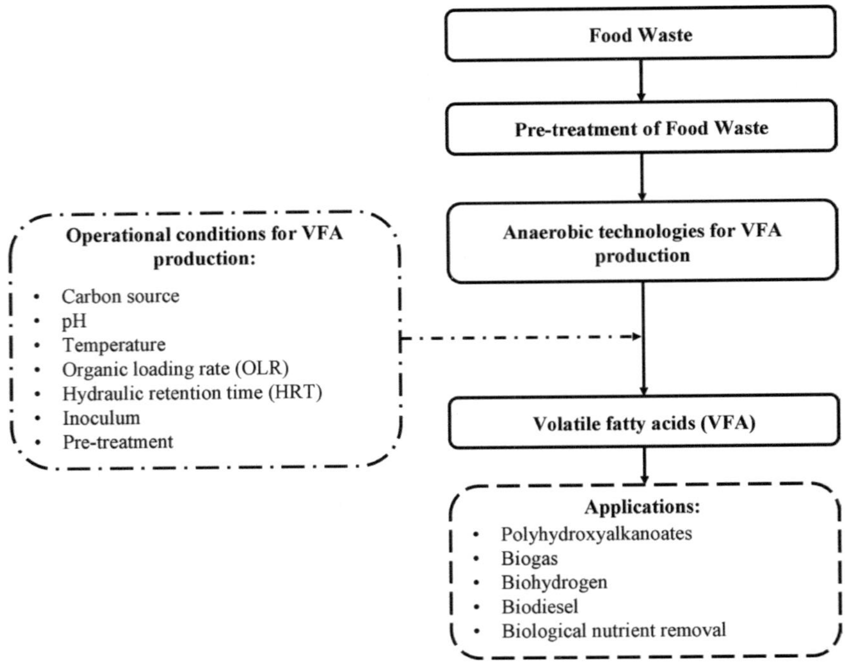

Figure 1.
Production of VFAs using FW as substrate: General process and applications overview. Adapted from lee et al. [2].

Actually, the production of VFAs is mainly accomplished by chemical routes through the oxidation or carboxylation of chemical precursors deriving from petroleum processing [6]. However, VFAs can also be biologically produced, using pure or mixed microbial cultures, in a single-stage anaerobic process. The use of mixed microbial cultures is emerging, as a broad spectrum of substrates can be used and sterile conditions are not required, lowering the production costs [7].

During the last years, several efforts have been done to improve the production of VFAs from FW through the assessment of different types of FW and optimization of operational conditions. Besides the type of FW used, operating parameters, such as pH, temperature, hydraulic retention time (HRT), inoculum, and organic loading rate (OLR) can affect the VFA production, composition and yield (**Figure 1**) [2, 3, 6]. VFAs have a broad spectrum of applications such as polyhydroxyalkanoates (PHAs), bioenergy (biogas, biohydrogen), biological nutrient removal, as well as in chemical industry as precursors in organic chemistry [2, 6]. Nowadays, it is known that by controlling the process it is possible to manipulate the VFA composition, which is an important factor considering the application of the VFA stream. For example, the manipulation of the VFA profile of the stream will allow to produce PHAs with different compositions and, consequently, with different applications (e.g. packaging, construction materials, medical applications, etc.) [8]. This chapter reviews sustainable processes for FW valorization through VFA production, which can minimize further environmental degradation and promote the evolution to a sustainable society, towards a circular economy.

2. Food waste

FW can be defined as "the final product of food chain that was not recycled or used for other purposes" [9] and corresponds to one third of the total food production for human consumption [10]. FW is one of the most produced waste and it is estimated to increase by 44% until 2025 due to, both, economy and population growth [10, 11]. This high increase in FW production have led to the need to develop appropriate treatment technologies [7]. Landfill, composting, incineration and animal feed are the conventional methods for FW disposal/treatment, which present several environmental concerns, such as air, soil and groundwater contamination, greenhouse gas emissions, odor production, leaching and disease propagation (in case of animal feed) [9, 10, 12]. As such, anaerobic digestion has been widely used as an eco-friendly, sustainable and low-cost alternative technology, that allows to treat the waste and valorize them by the recovery of added-value products, such as methane, hydrogen or VFAs [7, 10].

FW composition depends on the habits and economical level of the region and the climate, showing different characteristics, such as pH, solid content, and carbon to nitrogen ratio (C/N) [5, 10]. Notwithstanding, easy biodegradability, nutrients availability and moisture content are similar features worldwide [5]. FW is rich in carbohydrates (hemicellulose, cellulose, starch, and sugar like sucrose, fructose, and glucose), proteins, lipids and inorganic compounds [11, 12]. The sugar content varies between 35 and 60%, while proteins and lipids vary between 15–25% and 13–30%, respectively [10, 12]. Due to the high nitrogen content of proteins, FW presents a low C/N ratio comparing with other substrates. Moreover, FW have high content of other elements, as phosphorus, sodium, potassium, calcium or magnesium, and low content of trace elements, as iron, selenium, nickel or molybdenum [10]. All these features make the FW an interesting renewable source for VFA production. Different types of FW, such as solid waste of cafeteria [13, 14], tuna waste [15], fruit pulp waste [1, 16], cheese whey [8, 17], sugar cane molasses [17], corn stalk [18], potato peel waste [19], FW rich in proteins [20], brewers' spent grain [21], mixture of different fractions of FW [22], FW from canteen [23, 24], and vegetable wastes [25] have been used as feedstock in biological processes using mixed microbial cultures. Furthermore, FW can be also mixed with other wastes, like as waste activated sludge [26] and sewage sludge [27] to improve the biological process performance.

3. Production of volatile fatty acids (VFAs)

The conversion of the organic content of waste into VFAs requires an acidogenic fermentation (AF). While soluble organics can be directly fermented into organic acids and other fermentation products, such alcohols and hydrogen, insoluble compounds need to be hydrolyzed prior to acidification, limiting the rate of VFA production [28, 29]. As such, the production of VFAs involves two steps: (1) hydrolysis, and (2) acidogenesis, commonly occurring in a single anaerobic reactor (**Figure 2**) [2]. In the hydrolysis step, enzymes excreted by hydrolytic microorganisms (e.g. *Clostridium* sp., *Bacillus* sp., *Bifidobacterium* sp.) brake down complex organics (such as proteins, cellulose, lignin, and lipids) into simpler soluble monomers (such as amino acids, simple sugars, glycerol, and fatty acids), which lead to an increase in the soluble chemical oxygen demand (sCOD). Then, these monomers are mainly converted into VFAs (such as acetic, propionic, butyric, and valeric acids) by fermentative bacteria (e.g. *Acetovibrio cellulolytic*, *Butyrivibrio* sp., *Selenomonas* sp.) in the acidogenic fermentation step (acidogenesis) [2, 30–32].

The production of VFAs from acidogenic fermentation of FW involves a series of chemical reactions, where different metabolic pathways co-exist within the anaerobic digester. These pathways play a crucial role in the system performance and consequently in the FW conversion efficiency. Pyruvate is the primary intermediate

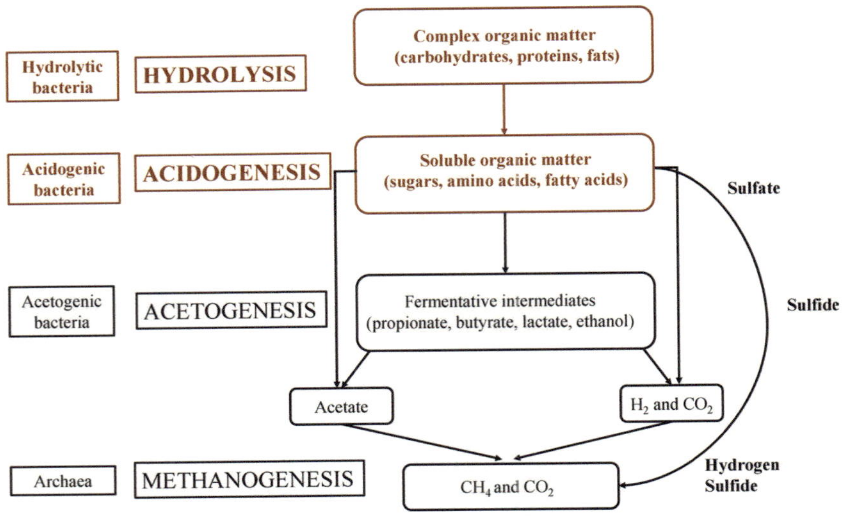

Figure 2.
Overall anaerobic digestion process schematic representation.

and can be converted into a wide range of products, such as VFAs, alcohols, hydrogen, and carbon dioxide. The type of substrate used, the environmental conditions and the microorganisms present in the reactor affects the proportions of pyruvate in each metabolic pathway and consequently the distribution of VFAs produced [5]. The acidogenic metabolic pathways can be classified in: acetate-ethanol type; propionate-type; butyrate-type; mixed-acid and lactate-type, depending on the main products produced during the acidogenic fermentation.

Acetate can be produced from acetyl-CoA pathway or from the syntrophic oxidation of ethanol or longer chain fatty acids. Ethanol can be produced from pyruvate in two or three steps, depending on the type of bacteria, and with acetyl-CoA and acetaldehyde as intermediates [5]. Propionate is produced by two distinct pathways: (1) pyruvate is reduced by the catalyzation of lactate dehydrogenase and then lactate is reduced to propionate through the propionate dehydrogenase; (2) propionate is produced by acidogenic bacteria (e.g., Corynebacteria, Propionibacterium and Bifidobacterium) via transcarboxylase cycle. Butyrate production from pyruvate comprises: (1) pyruvate conversion into acetyl-CoA by pyruvate dehydrogenase; (2) acetyl-CoA is converted into to butyryl-CoA with acetoacetyl-CoA, 3-hydroxybutyryl-CoA and crotonyl-CoA as intermediates sequentially by the catalysis of thiolase, 3- hydroxybutyryl-CoA dehydrogenase and butyryl-CoA dehydrogenase; (3) butyryl-CoA is converted into butyrate by phosphotransbutyrylase and butyrate-kinase enzymes or by the butyryl-CoA: acetate CoA-transferase [5]. For the lactate production, pyruvate is converted to lactate through lactate dehydrogenase and can be divide into two fermentation types: homolactate fermentation (one mole of glucose is converted into two moles of lactic acid) and heterolactate (lactic acid is produced with carbon dioxide and ethanol). In mixed fermentation, an equal amount of each acid is produced with a possible formation of carbon dioxide and hydrogen. This type of fermentation is common in FW fermentation with acetate and butyrate the main metabolites produced. Acetate can be also produced by homoacetogens, which are obligate anaerobes that can use hydrogen to reduce carbon dioxide to acetate. In autotrophic process, the homoacetogens consumed hydrogen and carbon dioxide producing acetate by Wood-Ljungdahl pathway [5].

Since acidogenesis is the second step of the anaerobic digestion of organic compounds into biogas (methane and carbon dioxide), high pH (above 8) or low pH (bellow 6), low temperature and/or low HRT are usually used to prevent methanogenic activity (**Figure 2**). Moreover, the operating parameters of the acidogenic fermentation, such as HRT, sludge retention time (SRT), organic loading rate (OLR), pH, temperature, and reactor configuration, must be optimized aiming at VFAs production yield maximization and at controlling the composition of the synthesized VFAs.

VFAs production from FW using mixed microbial cultures is mostly based on the use of suspended biomass. Thus, continuous stirred tank reactors (CSTR), stirred tank reactors (STR) and immersed membrane bioreactors are being applied for that purpose [1, 15, 21, 22, 33, 34]. Those reactors are usually operated in a continuous mode. However, for FW needing high retention times to be converted into VFAs, they are often converted into batch and semi-continuous (fed-batch) reactors [15, 19, 21, 33–38].

4. Effect of process parameters in VFAs production

The production of VFAs through FW fermentation using mixed microbial cultures, as well as the VFA composition and yield, are influenced by numerous factors, such as carbon source, pH, temperature, HRT, inoculum, and OLR [2, 3, 6, 7].

4.1 Carbon source

The type of carbon source, namely the complexity and composition, affects the acidification degree, since this depends on the readily fermentable organic fraction, and consequently, influences the production of VFAs [3, 25]. The content of carbohydrates, lipids and proteins influences the production of VFAs, since each fraction present different biodegradability and hydrolysis efficiency [6]. Moreover, the type of acids produced depends on the waste composition. The fermentation of carbohydrates, proteins and lipids form directly acetic, propionic, and butyric acids, while valeric and iso-valeric acids are related to the fermentation of proteins [15]. Lipids are the fraction more resistant to the biodegradation, being less suitable for AF than the other FW fractions, despite their high contribution for COD [6, 20]. For that reason, usually, lipids are separated and used to produce biodiesel instead of using to VFAs production [20].

Commonly, carbohydrates are easily converted into glucose and then fermented to VFAs [20]. However, the carbohydrate fraction, that is not readily degraded (cellulose, lignin or hemicellulose), can affect the VFA production rate, as well as the VFA concentration and yield (**Table 1**).

Zhang et al. [25] observed that high content of readily degradable carbohydrates promotes a faster production of VFAs, while the presence of not readily degraded fraction (e.g., cellulose) can delay the production of VFAs or even lead to a lower production. Another study, showed that different FW (cheese whey, sugarcane molasses and olive mill effluent) resulted in different VFAs production (**Table 1**) and also different acidification degree (cheese whey and sugarcane molasses: up to 40%; olive mill effluent: up to 12%) [39].

The type of carbon source and its composition (e.g. carbohydrates, lipids, proteins) lead to different acidogenic metabolic pathways and, consequently, different VFAs composition. Indeed, in Zhang's study [25], according to the VFA profile, different fermentation type were obtained (**Table 1**). Silva et al., [39] observed that for the readily fermentable wastes (cheese whey and sugarcane molasses), acetic

FW	Y_{VFA}/s	VFA concentration	Dominant VFA	Ref.
Potato peels	452 mgCOD/ gVS_{feed}	NA	Propionate (30.6%)	[25]
Carrots	321 mgCOD/gVS_{feed}	NA	Butyrate (54%)	[25]
Celery	372 mgCOD/gVS_{feed}	NA	Mixed acid	[25]
Chinese cabbage	201 mgCOD/ gVS_{feed}	NA	Acetic acid (~40%)	[25]
Cheese Whey	NA	3374 ± 138 mgCOD/L	Acetic acid (~50%)	[39]
Sugar cane molasses	NA	3110 ± 124 mgCOD/L	Acetic acid (~50%)	[39]
Olive mill effluent	NA	934 ± 113 mgCOD/L	Acetic acid (~70%)	[39]
Tofu	0.16 g/gVS	7.28 g/L	Acetic acid (56.3%)	[20]
Egg white	0.26 g/gVS	15.23 g/L	Mixed acid (~25% of each)	[20]

VS – volatile solids; NA – not available; Ref. - reference.

Table 1.
VFA production from FW using mixed microbial cultures.

acid was the main acid produced, followed by butyric acid (26–28%) and caproic and iso-valeric acids (10–14%), while for the waste with low acidogenic potential (olive mill effluent), acetic acid was followed by propionic acid (21–24%) and the production of heavier VFA (e.g., butyric, valeric acids) was not detected.

Proteins have a complex structure, making them less suitable to protease action and, consequently, a lower hydrolysis efficiency [6, 20]. Indeed, the hydrolysis of carbohydrates can be up to 80%, while protein hydrolysis is between 40% to 70%, which leads the latter to be considered the rate-limiting step in AF [6]. As carbohydrates, the origin of proteins affects the production and composition of VFAs. Shen et al., [20] obtained a high production of VFA using animal protein when compared with vegetal protein (**Table 1**). Independently of the protein, animal and vegetal, acetic, propionic, butyric and valeric acids were produced. However, the acids profile is different in the fermented vegetal and animal protein. This difference in composition can be related to the type of amino acids present in tofu and egg white. Even though it has been demonstrated that the carbon source affects the VFA production, FW feedstocks with similar compositions must still be individually investigated, as other factors, like operational conditions, should be considered.

4.2 pH

pH is one of the most important and critical parameters in VFA production, since it affects VFA concentration and composition due to its influence in hydrolysis and acidogenic process [3, 5]. pH affects the microorganisms activity since most of the enzymes cannot tolerate low or high pH environments (pH < 3 or pH > 12) [2, 5]. Moreover, pH can also promote the inhibition of methanogenic activity, since operating the reactor out of their optimal range (7.0–8.2), the methane production can be inhibited [33]. The optimal pH for VFA production depends on the type of waste and should fit both hydrolysis and acidogenic steps [2, 5]. Indeed, different wastes present different optimal pH for VFA production (**Table 2**).

FW	Operating conditions	Optimal pH	$[VFA]_{max}$	VFA composition HAc/HPr/HBut/ others (%)	Ref.
Tuna waste	Batch; 37 °C; RT 39 days	8	30.61 g COD/L	60/11/14/15	[15]
FW (canteen)	Batch; 35 °C; RT 10 days	9	10.9 g/L	60.5/4.6/12.8/22.1	[33]
Potato peel waste	CSTR batch; 37 °C; RT 5 days	7	41.9 g COD/L	46/29/24/1	[19]
FW (canteen and market)	semi-continuous reactor; 37 °C; HRT 2d	9	18.19 ± 0.72 g/L	37.9/12.3/46.8/3.0	[34]
FW and waste activated sludge	Batch; 20 °C; RT 12 days	8	8.24 gCOD/L	31.9/51.8/9.2/7	[26]

RT – retention time; HAc – acetic acid; HPr – propionic acid; HBut – butyric acid; others can include valeric acid, iso-butyric and iso-valeric acid, caproic acid or not VFA; Ref. - reference.

Table 2.
Optimal pH for VFA production from different carbon source.

Bermúdez-Penabad et al. [15] studied the effect of low (5 and 6), neutral (7) and alkaline (8, 9 and 10) pH in the acidogenic fermentation of tuna waste and obtained the highest VFA production under alkaline conditions, achieving the highest production at pH 8. Although, the highest hydrolysis was also obtained under alkaline conditions, indicating that more substrate is available for acidification, strong alkaline conditions (pH 10) seem to affect the acidogenic bacteria activity since the ratio VFA/sCOD was the lowest one. Acetic and butyric acids were the main VFAs produced (except for pH 10). The increase of pH from 7 to 10 led to a decrease of the butyric acid content, while the change of pH from 5 to 8 led to an increase of acetic acid. Another study, using FW from canteen as feedstock, also observed the highest VFA production at alkaline pH (9), being acetic acid the dominant acid (60.5%) [33]. Hussain et al., [14], using solid FW from cafeteria as feedstock, operated four thermophilic leach bed reactor in batch mode under pH 4–7 and observed an increase of VFA production with pH increase (from 6 gCOD/L to 36.5 gCOD/L, corresponding a maximum yield of 247 g COD/kg TVS_{added}). At all pH, acetic and butyric acids accounted to 80–85% of the total VFA, being acetic acid the main produced acid at pH 4 and 5 (55–61%) and butyric acid the dominant compound at pH 6 and 7 (48–54%). On the other hand, Ma et al., [38], using FW from cafeteria, obtained the highest production of VFA at pH 6 (53.87 g/L), with a significant production of propionic acid. The fermentation of potato peel waste showed a different trend of VFA composition, since butyric acid was the main compound at acidic (pH 5) and uncontrolled pH, while acetic acid was the main compound produced at alkaline and neutral pH [19]. Moreover, the highest production of VFA was achieved at pH 7 (41.9 gCOD/L and 0.63 gCOD/gVS_{fed}). Stein et al., [34] studied the effect of pH, temperature and HRT on the maximization of the butyric acid production. pH 9 led to the highest production of VFA, as well as the highest VFA yield (0.726 gVFA/gVS_{fed}), which was about 80% higher than the yield obtained at pH 7. Besides butyric acid, that was the dominant compound at pH 9, acetic acid was also produced as the dominant compound (**Table 2**) and was the dominant compound at pH 5.5 and 7 (34.8% and 47.9%, respectively). Propionic and iso-butyric acids were also produced. However, their content decreased at the highest pH tested. Wang et al., [23] studied the effect of pH (4, 5, 6 and uncontrolled) on the VFA production with two different inoculums (aerobic and anaerobic activated sludge). Regardless of the inoculum, VFA production was highest at pH 6, achieving a maximum of 30.8 gCOD/L (0.482 g/gVSS$_{removal}$, after 4 days of fermentation) and 51.3 gCOD/L (0.918 g/gVSS$_{removal}$, after 20 days of fermentation) for aerobic and anaerobic inoculum, respectively. Butyric acid was the dominant acid at pH 6, followed by acetic and propionic acids. Moreover, acetic and butyric acids were the predominant acids in all the pH tested, except for pH 4 and uncontrolled pH for aerobic inoculum, where acetic and propionic acids were the prevalent acids. Gouveia et al. [8] studied the impact of a dynamic variation of pH from 4 to 7, returning to pH 6 after each pH variation, in cheese whey derived-VFA production process. The production of organic acids was quite constant for all pH tested (about 13 gCOD/L), apart from pH 4 (about 4 gCOD/L). At pH 6, acetic acid was the main VFA, comprising 22–44%, and regardless of the dynamic variation of pH, the composition was always similar at pH 6. At high pH, the production of acetic acid was favored. Feng et al., [26] tested different pH (4–11) at room temperature and observed the highest concentration of VFAs at pH 8 (8.24 gCOD/L at 4 days of fermentation) during the co-digestion of FW with WAS. Acetic, propionic and butyric acids were the most prevalent VFA produced, achieving a total content of 82.5%, 91.9% and 92.9% at pH 5, 7 and 8, respectively. From pH 6 to 10, propionic acid was the main compound produced, while for pH 4 and 5, acetic acid was the main compound.

The pH effect on VFA production and profile using mixed cultures and FW does not present a direct relationship, being also dependent of the type of substrate used.

4.3 Temperature

Temperature is a key parameter that impacts the growth of microorganisms and their metabolism. Each microbial taxon has an optimal temperature range for its growth, and, therefore, a change on the operating temperature can affect the microbial population involved in the acidogenic fermentation [5, 6]. Mesophilic condition (25–45 °C) showed a similar or even higher VFA yield than thermophilic (50–60 °C) or hyperthermophilic (>65 °C) conditions [5]. On the other hand, thermophilic and hyperthermophilic conditions results in a higher hydrolysis and solubilization in comparison with mesophilic condition, which can lead to an increase of VFA production if an adequate microbial community is present [2, 13]. The composition of VFAs is also affected by the temperature, but in a less extent than that was observed for pH [2, 5]. The increase of temperature from mesophilic to thermophilic conditions leads to the metabolic shift from acetic acid to butyric acid [13]. Jiang et al. [36] reported that sCOD increased with temperature increase. However, the VFA concentration and yield at 55 °C was much lower than that obtained at 35 °C and 45 °C (14.90 g/L and 0.137 g/gVS$_{fed}$, 41.34 g/L and 0.379 g/gVS$_{fed}$, and 47.89 g/L and 0.440 g/gVS$_{fed}$, for 55, 35 and 45 °C, respectively), indicating a higher solubilization but a lower acidogenesis of FW at thermophilic condition (55 °C). The temperature increase resulted in a decrease of acetic and valeric acids content and an increase of butyric acid content. Acetic and propionic acid were the major compounds at 35 and 45 °C, representing ca. 70% of the total of VFAs. At 55 °C, butyric acid was the main compound, comprising more than 81% of the total VFAs and valeric acid was not detected. Although a higher ratio of VFA/COD and VFA concentration was observed at 45 °C, a high amount of energy is necessary to operate at this temperature. As such, 35 °C is considered as being the most cost-effective temperature for FW derived-VFA production process [36]. Similarly, He et al., [37] observed a decrease of VFA concentration, from 17 to 11 g/L, when the temperature increase from 35 to 55 °C. At 70 °C, the VFA production (about 13 g/L) was higher than 45 °C but lower than 35 °C. The hydrolysis rate was directly affected by temperature, having increased with temperature increase, indicating that higher temperatures promote the hydrolysis of FW. Comparing the three temperatures, acetic acid was the main compound at 70 °C, iso-butyric and butyric acids were the main compounds at 55 °C, while ethanol was the main compound at 35 °C, showing that the increase of temperature can inhibit ethanol production, favoring the production of VFAs, namely acetic and butyric acids. Zhang et al. [35], who have studied the effect of two temperatures (35 and 55 °C) at different pH (5, 6 and 7), reached the maximum VFA yield of 11.8 gCOD/L at pH 7 and 35 °C, being acetic, propionic and butyric acids the main produced VFAs (about 80% of total VFAs). For each initial pH applied, mesophilic conditions led to higher VFA concentration. When comparing all the conditions tested, except for pH 7 and 55 °C, thermophilic temperatures led to lower VFA production than mesophilic conditions. Applying the optimal conditions to a continuous reactor, an average VFA concentration and yield of 6.3 gCOD/L and 0.29 $gVFA/gVS_{added}$ was obtained.

Temperature (37, 55 and 70 °C) also affected the maximization of the butyric acid production [34]. Higher temperatures led to a decrease of VFA concentration, namely in butyric acid concentration, except at pH 7 and 55 °C, where an 280% increase of butyric acid concentration was achieved when compared to mesophilic conditions. The maximum concentration of butyric acid was achieved at pH 7 and 55 °C (10.55 g/L ± 0.17) and pH 9 and 37 °C (8.52 g/L ± 0.10).

Considering operating costs, mesophilic conditions (25–45 °C) are the most economical and efficient temperatures to produce VFAs.

4.4 Organic loading rate (OLR)

OLR corresponds to the amount of substrate, in this case FW, fed to the reactor per day and per unit of working volume and can be expressed in terms of COD, total solids (TS), volatile solids (VS), volatile suspended solids (VSS) or dissolved organic carbon (DOC) [2]. The OLR increase promotes the production of VFAs, since more substrate is available [5]. However, the presence of inhibiting substances in the FW is a key factor that must be considered when operating at high OLR, as might affect the hydrolytic and acidogenic bacteria, as well as methanogens [7]. Several studies on the effect of OLR on VFA production from FW are summarized in **Table 3**.

Carvalheira et al. [1] observed the increase of VFA production (from 19.9 ± 5.0 gCOD/L to 34.4 ± 5.0 gCOD/L) and the change of VFA composition with the increase of OLR (from 21.2 ± 3.2 gCOD/(L.d) to 51.1 ± 8.8 gCOD/(L.d)) at HRT of 1 day. Under the lowest OLR, acetic and propionic acids were the major compounds (19.65% and 19.11%, respectively), while under the highest OLR propionic acid was the prevailing VFA (10.07%) (**Table 3**). Similarly, Jiang et al. [36] also observed the increase of VFA concentration with the increase of OLR. However, it was observed a decrease of VFA after 12 days of operation, under the highest OLR, indicating that an OLR of 11 gTS/(L.d) is more appropriate for VFA production from FW. Acetic and butyric acids were the dominant compounds accounting 60–65% of the total VFAs in all OLR applied. Acetic and valeric acids increased with the OLR, while propionic and butyric acids decreased with the OLR (**Table 3**). Teixeira et al., [21] studied the effect of OLR on VFA production using raw brewers' spent grain as feedstock and observed an increase of VFA concentration in about 2.5 times (**Table 3**). Regardless of the OLR, no significant effects on the VFA composition was observed, being propionic acid the dominant acid produced (**Table 3**).

FW	Operating conditions	OLR	VFA production	VFA composition Hac/HPr/HBut/HVal (%)	Ref.
Peach pulp waste	CSTR; HRT 1 day; 30 °C; pH 5.5	21.2 ± 3.2 gCOD/(L.d)	19.9 ± 5.0 $gCOD_{FP}/L$	19.65/19.11/14.38/9.88	[1]
		33.3 ± 4.2 gCOD/(L.d)	24.0 ± 3.2 $gCOD_{FP}/L$	16.42/18.16/10.01/5.01	
		51.1 ± 8.8 gCOD/(L.d)	19.9 ± 5.0 $gCOD_{FP}/L$	5.48/10.07/7.76/7.14	
Simulated FW	Semi-continous; HRT 5 days; 35 °C; pH 6.0	5 gTS/(L.d)	13.27 g/L; 0.504 g/gVS_{fed}	27.46/23.57/33.26/15.71	[36]
		11 gTS/(L.d)	21.44 g/L; 0.411 g/gVS_{fed}	34.07/17.70/31.00/17.23	
		16 gTS/(L.d)	24.93 g/L; 0.306 g/gVS_{fed}	36.55/14.70/28.62/20.12	
Raw brewers' spent grain	STR; HRT 19 days; 30 °C; pH 4.7	4.3 gTS/(L.d)	15.8 ± 2.0 $gCOD_{FP}/L$	29.7/44.5/11.4/12	[21]
	STR; HRT 16 days; 30 °C; pH 4.7	16 gTS/(L.d)	35.5 ± 3.5 $gCOD_{FP}/L$	28.2/41.3/17.2/9.6	
Mixture of different fractions of food waste	IMB; HRT 5 days; 37 °C; pH 5.5 (R1)	4 gVS/(L.d)	11.44 ± 1.09 g/L; 0.51 g/VS_{added}	Hac/HPr/HBut/HCap 42.51/<5/19.06/27.46	[22]
	IMB; HRT 10 days; 37 °C; pH 5.5 (R1)	8 gVS/(L.d)	27.40 ± 2.33 g/L; 0.44 g/VS_{added}	Hac/HPr/HBut/HCap 18.47/8.64/47.54/15.72	
	IMB; HRT 5 days; 37 °C; pH 5.5 (R2)	6 gVS/(L.d)	16.04 ± 0.84 g/L; 0.52 g/VS_{added}	Hac/HPr/HBut/HCap 39.52/5.67/23.6/24.94	
	IMB; HRT 10 days; 37 °C; pH 5.5 (R2)	10 gVS/(L.d)	36.99 ± 1.68 g/L; 0.46 g/VS_{added}	Hac/HPr/HBut/HCap 25.80/8.9/37.43/12.95	

CSTR – continuous stirred tank reactor; STR – stirred tank reactor; IMB – immersed membrane bioreactor; FP – fermented products; Hcap – caproic acid; HRT – hydraulic retention time; NA – not available; Ref. – reference.

Table 3.
Effect of OLR on VFA production using FW as substrate.

From the fermentation of different fractions of FW, the main compounds were acetic, caproic and butyric acids (**Table 3**), observing an increase of butyric and propionic acid and a decrease of acetic and caproic acids with the increase of OLR (**Table 3**). Moreover, the high production of acetic and butyric acids can be related to the high OLR applied, since these acids have been related to reactor overloading during anaerobic digestion processes [22].

OLR also influences the composition of VFAs, however it has to be correlated with other factors, such as pH and HRT, which also affects the VFA composition [2].

4.5 Hydraulic retention time (HRT)

HRT can be described as the average length of time that the substrate and biomass remain inside the reactor [6]. HRT should be long enough to promote the hydrolysis and the acidogenic fermentation steps, which depends on the type of FW [6]. In theory, high HRT is advantageous for VFA production since the microbial population has more time to convert the substrate. However, a very high HRT reduce the quantity of waste to be treated per day and can favor the methanogens activity, if suitable pH is applied [5, 6]. Moreover, high HRT can lead to VFA yield stabilization due to feedstock limitation [5]. The optimal HRT can vary even for the same feedstock [7]. Teixeira et al., [21] studied the effect of HRT (19 and 41 days) on the production of VFA at OLR of 4.3 gTS/(L.d). Due to the kind of substrate (solid and complex substrate without pre-treatment), high HRT was applied. The increase of HRT boosted the production of VFA (from 11.2 gFP/L to 24.4 gFP/L), since a longer contact between microorganisms and substrate was promoted. The prevailing acid produced was propionic acid, followed by acetic, butyric and valeric acids. However, their content was similar for both tested HRT, showing that HRT had no significant impact on the VFA composition. In another study, using a mixture of sewage sludge with cheese whey as feedstock, a sequential increase of HRT (from 10 to 20 days) was investigated [27]. The change of HRT increased the acidification degree (from 27 to 45%) with a similar ratio of VFA/sCOD (85 and 89%). The increase of HRT also promoted a slightly change on the VFA composition, with a slightly increase of iso-butyric and butyric acids (from 51 to 55%) and a decrease of acetic acid (from 33 to 24%). HRT depends not only on the type of substrate, but also on other operational parameters. As most of the studies are performed in batch reactors, the information about HRT effect on the VFA production is scarce.

4.6 Inoculum

The type of microorganisms present in the mixed microbial cultures may affect the acids production. It is necessary a careful selection of the microbial population present in the acidogenic fermentation process. If inadequate, a disparity on microbial populations can delay or limit the fermentative reactions and pathways, lowering the process yields [13]. Anaerobic inoculum from anaerobic sludge digesters obtained higher FW hydrolysis and VFA yield than aerobic inoculum from activated sludge process [13]. Atasoy et al. [40] investigated the effect of three types of inoculum with two different physical sludge structure (small and large granular sludge and anaerobic digester sludge) on VFAs production and composition. The highest VFA production was obtained with large granular sludge (1.99 ± 0.06 gCOD/L), followed by anaerobic digester sludge (1.14 ± 0.07 gCOD/L) and small granular sludge (1.06 ± 0.12 gCOD/L). As for VFA production, large granular sludge led to the highest VFA yield (0.97 gVFA/g_SCOD). For small granular and anaerobic digester sludge a similar yield was obtained (0.36 and 0.38 gVFA/g_SCOD, respectively), indicating that VFA production efficiency changed with the inoculum type.

Moreover, this study showed the ability of granular sludge to attain high VFA production efficiencies instead of biogas production, which is the usual application for granular sludge. Butyric and propionic acids were the main VFAs produced using large and small granular sludge, respectively, while acetic and propionic acids had similar content with anaerobic digester sludge. Wang et al. [23] studied the effect of pH (4.0, 5.0, 6.0 and not controlled) using two types of inoculum (anaerobic and aerobic). The highest VFA concentrations were obtained at pH 5.0 and 6.0, independently of the inoculum. Moreover, when anaerobic inoculum was used, it was obtained a slightly higher VFA production when compared with aerobic inoculum, which could be related to a higher acidogenic bacteria content present in the anaerobic inoculum and higher microbial activities under anaerobic conditions. Acetic and butyric acids were the main compounds (representing 90% of total VFAs) in all experiments except at pH 4.0 and not controlled pH using aerobic inoculum, where acetic and propionic acids were the major acids produced. Another study [13] evaluated the effect of inoculum source (mesophilic anaerobic sludge from biosolids digester of a municipal wastewater treatment plant (35 °C), thermophilic anaerobic sludge treating flour residues (55 °C) and hyperthermophilic anaerobic sludge treating microalgae) on the production of VFAs under mesophilic, thermophilic and hyperthermophilic conditions (70 °C). The mesophilic and thermophilic reactors were also operated at 70 °C. The hydrolysis efficiency was similar for all reactors, ranging between 27 and 40%. The reactor operation obtained under thermophilic conditions led to the highest fermentation yield (0.44 gCOD/gVSS-COD_{added}), followed by mesophilic (0.33 gCOD/gVSS-COD_{added}) and hyperthermophilic conditions (0.08 gCOD/gVSS-COD_{added}). Moreover, the fermentation yield at 70 °C using mesophilic and thermophilic were lower than that obtained at standard conditions (0.30 gCOD/gVSS-COD_{added} and 0.28 gCOD/gVSS-COD_{added}, respectively). VFAs accounted to ca. 60–71% of the solubilized matter at mesophilic and thermophilic conditions, with acetic acid the major compound (70%) at mesophilic temperature and butyric acid (60%) the major compound at thermophilic temperature. The higher production of VFA at 35 °C and 55 °C revealed the importance of inoculum source in the improvement of acidogenic activity. Thus, different inoculum types present a variability in microbial populations which lead to a distinct performance of hydrolytic and acidogenic processes.

4.7 Pre-treatment

The pre-treatment will make the complex compound of FW more accessible for the hydrolysis, being the rate-limiting step of fermentation process.

Pre-treatments can be divided in physical, chemical and biological categories [41]. Physical pre-treatment present the high efficiency in terms of degradation but present high costs related to high energy consumption. Chemical pre-treatment is a cheap and efficient process but is not environmentally appealing and the chemicals used may cause fermentation inhibition. Biological pre-treatment presents several advantages (natural process, environmentally friendly, non-toxic for fermentation, economic) but are slower than physical and chemical processes.

Physical pre-treatment comprises heat, mechanical and radiation processes to change the structure and/or composition of FW. In terms of chemical pre-treatment the most commonly used are alkaline or acidic solutions, hydrogen peroxide and ozone. Alkaline and acidic pre-treatments break the cell wall promoting the solubilization. However, these pre-treatments may cause equipment corrosion and interfere with the fermentation pH. Ozone pre-treatment is safer but expensive. Hydrogen peroxide is toxic for the environment and causes cell growth inhibition, even though it presents a high solubilization degree [41]. Biological pre-treatment comprises microbial (inoculated microorganisms as single culture or consortia and microorganisms from matured compost) and enzymatic (single or mixed enzymes) pre-treatment. Enzymatic pre-treatment is faster than microbial pre-treatment. However, it can be a costly treatment due to the operating costs involved in the production and extraction processes [41].

Guo et al., [18] evaluated the effect of different pre-treatments (sulfuric acid, acetic acid, aqueous ammonia, sodium hydroxide and steam explosion) of corn stalk on the production of organic acids. Steam explosion was the most suitable process for microbial growth and VFAs production, achieving a total of 2.98 g/L. The lowest production of VFAs was achieved using the acetic acid pre-treatment. Shen et al. [20] investigated the effect of hydrothermal pre-treatment (160 °C and 30 minutes), on tofu and egg white and assessed the production of VFAs using the pre-treated and untreated feedstock, observing that the pre-treatment improved the VFA production from tofu but not from egg white. Treated tofu reached a maximum VFA concentration of 21.07 g/L and a yield of 0.46 g/gVS, while with treated egg white a maximum VFA concentration of 11.45 g/L and a VFA yield of 0.20 g/gVS were achieved. Contrarily to the VFA yields, the VFA composition was not affected by the hydrothermal treatment. Yin et al., [24] also assessed the effect of hydrothermal pre-treatment (140, 160, 180 and 200 °C and 30 minutes) on the production of VFAs from FW. The hydrothermal pre-treatment of FW enhanced the VFA production. The optimal hydrothermal temperature was at 160 °C, where a VFA yield of 0.908 $g/gVS_{removal}$ and a VFA concentration of 34.1 g/L (increase of 47.6% compared with control) was reached. Independently of pre-treatment temperature, butyric and acetic acids (between 38.6–41.2% and 31.1–35.2%, respectively) were the dominant acids, followed by propionic acid (about 20%) and valeric acid (about 8%, except for pre-treatment at 200 °C (14.51%)). Pre-treatment of FW is one option to achieve high solubilization, improving the FW biodegradability and consequently enhance the VFAs production.

4.8 Other factors

Other factors, such as total solids (TS) [15, 42], co-digestion [43] and substrate shift [16, 17] also affect the VFA production.

The initial total solid (TS) concentration can limit the mass transfer between the substrate and the microorganisms [42]. Wang et al., [42] tested four initial TS concentrations (40, 70, 100 and 130 g/L) observing a slower VFAs production at higher TS concentration, although a higher VFA concentration was obtained (62.24 gCOD/L). The increase of TS content, from 40 to 130 g/L, led to a maximum VFA concentration of 26.10, 39.68, 59.58 and 62.64 gCOD/L. On the other hand, the increase of TS content led to a decrease of VFA yield (0.799, 0.644, 0.604 and 0.467 $gCOD/gVS_{fed}$) and acidification degree (48.2%, 42.7%, 41.2%, and 35.8%). Propionic acid accounted to 30.19–34.86% of the total VFAs and was not affected by the TS concentration, while a higher content of butyric acid and a lower content of acetic acid were achieved at higher TS content. Bermúdez-Penabad et al., [15], also observed an increase of VFA concentration with the increase of TS content (from 2.5 to 8%TS (w/v)) and obtained the higher VFA yield and acidification at lower TS concentration (0.73 $gCOD_{VFA}/gCOD_{waste}$ and 73%, respectively). The lower yields at higher TS content could be related to inhibition at high VFA concentration. Independently of TS content, acetic acid was the dominant acid produced, followed by butyric, iso-valeric and propionic acids.

Co-digestion consists in the simultaneous treatment of two or more substrates. Although the mono-digestion of FW is suitable, co-digestion presents several advantages and benefits, like as an improvement of nutrients balance, synergistic effects between microorganisms, dilution of potential toxicity, increase of digestion rate [10, 26]. FW can be mixed with different other wastes to improve the system performance [10]. In Feng's study [26] an improvement in VFA production was obtained by the addition of FW to waste activated sludge fermentation. The production of VFA with only waste activated sludge or FW was 971.7 and 1468.5 mgCOD/L, respectively, while in the co-digestion increased to 8236.6 mgCOD/L. Moreover, the VFA composition obtained from waste activated sludge or FW fermentation was different than that obtained in the co-digestion, although acetic and propionic acids were the prevalent acids produced. Another study [27] also showed that the increase of cheese whey content on the sewage sludge digestion increased the production of VFAs. Using only sewage sludge as substrate the maximum concentration obtained was 1507 mgCOD/L, while the addition of cheese whey, in a ratio of 25:75, allowed to increase the VFA concentration up to 3226 mgCOD/L. The addition of cheese whey also led to a changed in the VFA profile, with acetic, propionic and butyric acids being the main acids produced, while acetic, propionic and iso-valeric acids were the prevalent acids produced in sewage sludge fermentation.

Most of the wastes are seasonal which can affect the continuous production of the VFAs. So, two possible solutions may be used: (1) feedstock storage or (2) feedstock shift. In the first option, the storage might require huge buffer tanks or facilities, besides the possible feedstock degradation during the storage. In the second option, the change between different feedstock could led to a different VFA composition and production, besides could affect the robustness of the continuous operation. Duque et al., [17] studied the feedstock shift from cheese whey to sugar cane molasses to cheese whey, observing an immediate response to the feedstock shift by the change of fermented products profile. The highest fermented products concentration (13.2 gCOD/L) was achieved using sugar cane molasses, being propionic and valeric acids the most prevalent compounds. During the operation with cheese whey, a maximum concentration of fermented products of 9.7–10.6 gCOD/L was achieved, being acetic and butyric acids the dominant compounds. Although the feedstock shift from cheese whey to sugar cane molasses changed the fermented products concentration and composition as well as the kinetic parameters, the shift from SCM to cheese whey demonstrated the process robustness since the system responded similarly to the first cheese whey. Mateus et al., [16] assessed the effect of the feedstock shift and operational conditions (pH and OLR/HRT) of three pulp waste (peach, raspberry and white guava) on acidogenesis. The authors observed that, independently of pulp composition, the fermented products profiles were similar and stabilized over a short period of time after each operation change (feedstock shift or operational conditions), showing the robustness of the system. Butyric, acetic and valeric acids were the main acids produced in all the conditions tested. Generally, the latter studies showed the ability of the microbial community to deal with FW feedstock shift with no need to stop the operation, representing an important advantage at full scale operation.

5. Applications of VFAs

VFAs are valuable products with a huge market demand and a wide range of applications such as polyhydroxyalkanoates, biodiesel, biogas, biohydrogen, and biological nutrient removal [2, 3, 5].

5.1 Polyhydroxyalkanoates (PHAs)

Polyhydroxyalkanoates (PHAs) are thermoplastic biodegradable polyesters produced by microorganisms from renewable resources, such as VFAs [2, 29].

PHA has a wide range of applications, such as packaging, compost bags, agriculture/horticulture films, durable and consumer retail goods [29] and besides to be environmental-friendly, their implementation has been limited due to the high production cost when compared with the conventional plastics [2, 17]. Usually, industrial PHA production use pure cultures and expensive substrates (e.g., sucrose or glucose), so it is expected that the combination of mixed microbial cultures, which do not require sterile conditions, and low-cost substrates, such as VFAs from FW, will contribute for the decrease of operational costs due to reduction of substrate cost and saving energy [17, 29]. PHA production from mixed microbial cultures comprises three-stages: (1) acidogenic fermentation, where the organic matter is converted into VFAs; (2) selection of mixed microbial cultures, where the microbial culture is enriched in PHA accumulating organism; (3) PHA production, where the mixed microbial cultures selected in the second stage is fed with the VFAs produced in the first stage at the culture's maximum PHA accumulation [3, 17]. Using this process, a PHA content of 40–77% can be achieved from fermented FW [2]. The VFA composition establishes the PHA composition, which defines the physical and mechanical properties of the polymer. Acetic and butyric acids are converted into polyhydroxybutyrate (PHB), while propionic and valeric acids are converted into polyhydroxyvalerate (PHV). PHB is the most common PHA produced, however due to their properties (brittle and stiff), present limited applications [2, 29]. The incorporation of HV monomers result in a co-polymer (P(HB-co-HV)), which is more flexible and tougher [2].

5.2 Bioenergy

The increase of energy demand, as well as the depletion of oil reserves have been led to the development of suitable alternatives for energy resources. Waste-derived VFA is a low-cost source for the generation of different types of energy, such as biogas, biohydrogen or biodiesel (valuable fuels).

5.2.1 Biogas

Biogas, the final product of anaerobic digestion, is mainly composed by methane so it can be used as green energy source (energy value of 37.38 kJ/L) for heat and power generation [16]. The most anaerobic digestion processes use a single reactor, where VFAs are the intermediate product. However, acidogens and methanogens are not subjected to their optimal growth and activity conditions, which can affect the system performance [1, 2]. In order to provide the optimal conditions for the microorganisms and avoid the methanogens inhibition, due to a quick acidification of FW, a two-stage system can be used. In this case, the hydrolytic/acidogenic stage is separated from the methanogenic stage [1, 16]. Hydrolytic/acidogenic stage can be operated at acidic pH and low HRT, producing VFAs and hydrogen, while the methanogenic stage can be operated at neutral pH and high HRT, producing biogas rich in methane from the VFAs obtained in the first stage [1, 2]. Among all the VFAs, propionic acid is the main acid that can negatively impact the methanogenic activity, and consequently the biogas production, at concentrations above 1.36–2.27 gCOD/L [16].

5.2.2 Biohydrogen

Hydrogen is considered the future fuel and one of the most attractive renewable energy due to their efficiency characteristics [6]. Biohydrogen can be produced in the first stage of the two-stage anaerobic digestion process and can be used as a renewable energy source (energy value of 12.71 kJ/L), boosting the energy recovery of the process [16]. Biohydrogen production potential is affected by the content on carbohydrates, as these are the preferred substrates for hydrogen production [16].

Biohydrogen can be also produced by photo fermentation, where the conversion of VFAs into hydrogen is performed by purple non-sulfur bacteria in the presence of light and using several organic compounds as feedstock [44]. Inhibitory compounds of FW, temperature, pH, wastewater color, light intensity and wavelength can affect the hydrogen production [44]. Moreover, the type of carbon also affects the efficiency of the process, due to the variation in electron transfer capabilities in different metabolic pathways [44].

5.2.3 Biodiesel

Biodiesel is a methyl ester of long-chain fatty acids, which can be obtained from lipids through transesterification process. Biodiesel is a renewable energy source, however its production presents high costs due to the use of costly raw materials (about 70–75% of the total cost) [2]. Therefore, the production of biodiesel from waste-derived VFA have been gaining attention, where the VFAs can be converted into microbial lipids for biodiesel production [3]. The utilization of VFA as feedstock resulted in high lipids production and yield, showing the ability of VFA to produce lipids. VFA composition, pH, temperature, strains, inoculum concentration and nitrogen to carbon ratio can impact the lipids production [3].

5.3 Biological nutrient removal

Biological nitrogen removal is a biological process that includes aerobic nitrification followed by anoxic denitrification for the nitrogen removal. It is known that VFAs are an important carbon substrate for nitrogen removal, representing an economical alternative as feedstock for denitrification. Moreover, phosphorus can be removed by enhanced biological phosphorus removal process, where the microorganisms are subjected to anaerobic and aerobic conditions and VFAs are used as carbon source. Alternating between anaerobic-aerobic-anoxic conditions, a simultaneous nitrogen and phosphorus removal can be achieved [2, 3]. The denitrification efficiency and rate can be influenced by the composition of VFA stream, being acetic and propionic acids the preferred acids due to their high nitrate removal rates [3]. On the other hand, propionic acid present a high phosphorus removal efficiency [2]. Besides the type of VFA, other operational parameters, such as type of reactor/process, dissolved oxygen, temperature, and pH can affects the efficiency of biological nutrient removal process.

6. Conclusion(s)

The biological production of VFAs from FW is one of the most promising sources for resource recovery due to the high availability of this waste. The production and recovery of VFAs from waste will accomplish one of the goals for this century, allowing to reduce the consumption of raw materials, as well as the waste accumulation/pollution and is in line with the circular economy approach by the substitution of the linear economic model of "take-make-consume-dispose" by the circular economy model. It has been demonstrated that VFAs have a great potential to be used for several applications, such as PHA and bioenergy production. Several process conditions (e.g. type of FW, pH, HRT, OLR, temperature, etc.) affects the VFA production in terms of VFA profile, concentration, and yield. Thus, these factors must be taken into consideration when implementing this process to fit the final application. For instance, the acidogenic fermentation can be manipulated in order to produce different VFAs profiles, leading to PHAs with different compositions (taylor-made). At last, several pre-treatments can be applied to FW, enhancing the acidification process. However, positive and negative aspects must be considered when choosing the type of pre-treatment, namely costs and environmental concerns.

Acknowledgements

The authors grateful acknowledge Fundação para a Ciência e a Tecnologia (FCT) for the financial support through the project PTDC/BTA-BTA/31746/2017. This work was supported by the Applied Molecular Biosciences Unit- UCIBIO which is financed by national funds from FCT/MCTES (UID/Multi/04378/2019).

Conflict of interest

The authors declare no conflict of interest.

Author details

Mónica Carvalheira and Anouk F. Duque*
UCIBIO-REQUIMTE, NOVA School of Science and Technology, NOVA University of Lisbon, Caparica, Portugal

*Address all correspondence to: af.duque@fct.unl.pt

References

[1] Carvalheira M, Cassidy J, Ribeiro JM, Oliveira BA, Freitas EB, Roca C, et al. Performance of a two-stage anaerobic digestion system treating fruit pulp waste: The impact of substrate shift and operational conditions. Waste Management. 2018;**78**:434-445. DOI: 10.1016/j.wasman.2018.06.013

[2] Lee WS, Chua ASM, Yeoh HK, Ngoh GC. A review of the production and applications of waste-derived volatile fatty acids. Chemical Engineering Journal. 2014;**235**:83-99. DOI: 10.1016/j.cej.2013.09.002

[3] Atasoy M, Owusu-Agyeman I, Plaza E, Cetecioglu Z. Bio-based volatile fatty acid production and recovery from waste streams: Current status and future challenges. Bioresource Technology. 2018;**268**:773-86. DOI: 10.1016/J.BIORTECH.2018.07.042

[4] Ravindran R, Jaiswal AK. Exploitation of Food Industry Waste for High-Value Products. Trends in Biotechnology. 2016;**34**(1):58-69. DOI: 10.1016/j.tibtech.2015.10.008

[5] Zhou M, Yan B, Wong JWC, Zhang Y. Enhanced volatile fatty acids production from anaerobic fermentation of food waste: A mini-review focusing on acidogenic metabolic pathways. Bioresource Technology. 2018;**248**:68-78. DOI: 10.1016/j.biortech.2017.06.121

[6] Strazzera G, Battista F, Garcia NH, Frison N, Bolzonella D. Volatile fatty acids production from food wastes for biorefinery platforms: A review. Journal of Environmental Management. 2018;**226**:278-88. DOI: 10.1016/j.jenvman.2018.08.039

[7] Wainaina S, Lukitawesa, Kumar Awasthi M, Taherzadeh MJ. Bioengineering of anaerobic digestion for volatile fatty acids, hydrogen or methane production: A critical review. Bioengineered. 2019;**10**(1):437-58. DOI: 10.1080/21655979.2019.1673937

[8] Gouveia AR, Freitas EB, Galinha CF, Carvalho G, Duque AF, Reis MAM. Dynamic change of pH in acidogenic fermentation of cheese whey towards polyhydroxyalkanoates production: Impact on performance and microbial population. New Biotechnology. 2017;**37**:108-16. DOI: 10.1016/j.nbt.2016.07.001

[9] Lin CSK, Pfaltzgraff LA, Herrero-Davila L, Mubofu EB, Abderrahim S, Clark JH, et al. Food waste as a valuable resource for the production of chemicals, materials and fuels. Current situation and global perspective. Energy and Environmental Science. 2013;**6**(2):426-64. DOI: 10.1039/c2ee23440h

[10] Capson-Tojo G, Rouez M, Crest M, Steyer JP, Delgenès JP, Escudié R. Food waste valorization via anaerobic processes: a review. Reviews in Environmental Science and Biotechnology. 2016;**15**(3):499-547. DOI: 10.1007/s11157-016-9405-y

[11] Dahiya S, Kumar AN, Shanthi Sravan J, Chatterjee S, Sarkar O, Mohan SV. Food waste biorefinery: Sustainable strategy for circular bioeconomy. Bioresource Technology. 2018;**248**:2-12. DOI: 10.1016/j.biortech.2017.07.176

[12] Waqas M, Rehan M, Khan MD, Nizami AS. Conversion of food waste to fermentation products. In: Encyclopedia of Food Security and Sustainability. Elsevier; 2018. p. 501-9. DOI: 10.1016/B978-0-08-100596-5.22294-4

[13] Arras W, Hussain A, Hausler R, Guiot SR. Mesophilic, thermophilic and hyperthermophilic acidogenic fermentation of food waste in batch:

Effect of inoculum source. Waste Management. 2019;**87**:279-87. DOI: 10.1016/j.wasman.2019.02.011

[14] Hussain A, Filiatrault M, Guiot SR. Acidogenic digestion of food waste in a thermophilic leach bed reactor: Effect of pH and leachate recirculation rate on hydrolysis and volatile fatty acid production. Bioresource Technology. 2017;**245**:1-9. DOI: 10.1016/j.biortech.2017.08.130

[15] Bermúdez-Penabad N, Kennes C, Veiga MC. Anaerobic digestion of tuna waste for the production of volatile fatty acids. Waste Management. 2017;**68**:96-102. DOI: http://dx.doi.org/10.1016/j.wasman.2017.06.010

[16] Mateus S, Carvalheira M, Cassidy J, Freitas E, Oehmen A, Reis MAM. Two-stage anaerobic digestion system treating different seasonal fruit pulp wastes: Impact on biogas and hydrogen production and total energy recovery potential. Biomass and Bioenergy. 2020;**141**:105694. DOI: 10.1016/j.biombioe.2020.105694

[17] Duque AF, Oliveira CSS, Carmo ITD, Gouveia AR, Pardelha F, Ramos AM, et al. Response of a three-stage process for PHA production by mixed microbial cultures to feedstock shift: Impact on polymer composition. New Biotechnology. 2014;**31**(4):276-88. DOI: 10.1016/j.nbt.2013.10.010

[18] Guo P, Mochidzuki K, Cheng W, Zhou M, Gao H, Zheng D, et al. Effects of different pretreatment strategies on corn stalk acidogenic fermentation using a microbial consortium. Bioresource Technology. 2011;**102**(16):7526-31. DOI: 10.1016/j.biortech.2011.04.083

[19] Lu Y, Zhang Q, Wang X, Zhou X, Zhu J. Effect of pH on volatile fatty acid production from anaerobic digestion of potato peel waste. Bioresource Technology. 2020;**316**:123851. DOI: 10.1016/j.biortech.2020.123851

[20] Shen D, Yin J, Yu X, Wang M, Long Y, Shentu J, et al. Acidogenic fermentation characteristics of different types of protein-rich substrates in food waste to produce volatile fatty acids. Bioresource Technology. 2017;**227**:125-32. DOI: 10.1016/j.biortech.2016.12.048

[21] Ribau Teixeira M, Guarda EC, Freitas EB, Galinha CF, Duque AF, Reis MAM. Valorization of raw brewers' spent grain through the production of volatile fatty acids. New Biotechnology. 2020;**57**:4-10. DOI: 10.1016/J.NBT.2020.01.007

[22] Wainaina S, Awasthi MK, Horváth IS, Taherzadeh MJ. Anaerobic digestion of food waste to volatile fatty acids and hydrogen at high organic loading rates in immersed membrane bioreactors. Renewable Energy. 2020;**152**:1140-8. DOI: 10.1016/j.renene.2020.01.138

[23] Wang K, Yin J, Shen D, Li N. Anaerobic digestion of food waste for volatile fatty acids (VFAs) production with different types of inoculum: Effect of pH. Bioresource Technology. 2014;**161**:395-401. DOI: 10.1016/j.biortech.2014.03.088

[24] Yin J, Wang K, Yang Y, Shen D, Wang M, Mo H. Improving production of volatile fatty acids from food waste fermentation by hydrothermal pretreatment. Bioresource Technology. 2014;**171**:323-9. DOI: 10.1016/j.biortech.2014.08.062

[25] Zhang Q, Lu Y, Zhou X, Wang X, Zhu J. Effect of different vegetable wastes on the performance of volatile fatty acids production by anaerobic fermentation. Science of the Total Environment. 2020;**748**:142390. DOI: 10.1016/j.scitotenv.2020.142390

[26] Feng L, Yan Y, Chen Y. Co-fermentation of waste activated sludge with food waste for short-chain fatty acids production: Effect of pH

at ambient temperature. Frontiers of Environmental Science and Engineering in China. 2011;**5**(4):623-32. DOI: 10.1007/s11783-011-0334-2

[27] Iglesias-Iglesias R, Kennes C, Veiga MC. Valorization of sewage sludge in co-digestion with cheese whey to produce volatile fatty acids. Waste Management. 2020;**118**:541-51. DOI: 10.1016/j.wasman.2020.09.002

[28] Bengtsson S, Hallquist J, Werker A, Welander T. Acidogenic fermentation of industrial wastewaters: effects of chemostat retention time and pH on volatile fatty acids production. Biochemical Engineering Journal. 2008;**40**(3):492-9. DOI: 10.1016/j.bej.2008.02.004

[29] Reis M, Albuquerque M, Villano M, Majone M. Mixed culture processes for polyhydroxyalkanoate production from agro-industrial surplus/wastes as feedstocks. In: Encyclopaedia Comprehensive Biotechnology. 2nd ed. Elsevier B.V.; 2011. p. 669-84. DOI: 10.1016/B978-0-08-088504-9.00464-5

[30] Chen JL, Ortiz R, Steele TWJJ, Stuckey DC. Toxicants inhibiting anaerobic digestion: A review. Biotechnology Advances. 2014;**32**(8):1523-34. DOI: 10.1016/j.biotechadv.2014.10.005

[31] Lin L, Li X yan. Acidogenic fermentation of iron-enhanced primary sedimentation sludge under different pH conditions for production of volatile fatty acids. Chemosphere. 2018;**194**:692-700. DOI: 10.1016/j.chemosphere.2017.12.024

[32] Veluchamy C, Kalamdhad AS. Biochemical methane potential test for pulp and paper mill sludge with different food/microorganisms ratios and its kinetics. International Biodeterioration & Biodegradation. 2017;**117**:197-204. DOI: 10.1016/J.IBIOD.2017.01.005

[33] Cheah YK, Vidal-Antich C, Dosta J, Mata-Álvarez J. Volatile fatty acid production from mesophilic acidogenic fermentation of organic fraction of municipal solid waste and food waste under acidic and alkaline pH. Environmental Science and Pollution Research. 2019;**26**(35):35509-22. DOI: 10.1007/s11356-019-05394-6

[34] Stein UH, Wimmer B, Ortner M, Fuchs W, Bochmann G. Maximizing the production of butyric acid from food waste as a precursor for ABE-fermentation. Science of the Total Environment. 2017;**598**:993-1000. DOI: 10.1016/j.scitotenv.2017.04.139

[35] Zhang L, Loh KC, Dai Y, Tong YW. Acidogenic fermentation of food waste for production of volatile fatty acids: Bacterial community analysis and semi-continuous operation. Waste Management. 2020;**109**:75-84. DOI: 10.1016/j.wasman.2020.04.052

[36] Jiang J, Zhang Y, Li K, Wang Q, Gong C, Li M. Volatile fatty acids production from food waste: Effects of pH, temperature, and organic loading rate. Bioresource Technology. 2013;**143**:525-30. DOI: 10.1016/j.biortech.2013.06.025

[37] He M, Sun Y, Zou D, Yuan H, Zhu B, Li X, et al. Influence of Temperature on Hydrolysis Acidification of Food Waste. Procedia Environmental Sciences. 2012;**16**:85-94. DOI: 10.1016/j.proenv.2012.10.012

[38] Ma J, Xie S, Yu L, Zhen Y, Zhao Q, Frear C, et al. pH shaped kinetic characteristics and microbial community of food waste hydrolysis and acidification. Biochemical Engineering Journal. 2019;**146**:52-9. DOI: 10.1016/j.bej.2019.03.004

[39] Silva FC, Serafim LS, Nadais H, Arroja L, Capela I. Acidogenic fermentation towards valorisation of organic waste streams into

volatile fatty acids. Chemical and Biochemical Engineering Quarterly. 2013;**27**(4):467-76.

[40] Atasoy M, Eyice O, Cetecioglu Z. A comprehensive study of volatile fatty acids production from batch reactor to anaerobic sequencing batch reactor by using cheese processing wastewater. Bioresource Technology. 2020;**311**:123529. DOI: 10.1016/j.biortech.2020.123529

[41] Abd-Aziz S, Ibrahim MF, Jenol MA. Biological Pretreatment of Lignocellulosic Biomass for Volatile Fatty Acid Production. In: Emerging Areas in Bioengineering. Weinheim, Germany: Wiley-VCH Verlag GmbH & Co. KGaA; 2018. p. 191-201. DOI: 10.1002/9783527803293.ch11

[42] Wang Q, Jiang J, Zhang Y, Li K. Effect of initial total solids concentration on volatile fatty acid production from food waste during anaerobic acidification. Environmental Technology (United Kingdom). 2015;**36**(15):1884-91. DOI: 10.1080/09593330.2015.1015454

[43] Paudel S, Kang Y, Yoo YS, Seo GT. Effect of volumetric organic loading rate (OLR) on H2 and CH4 production by two-stage anaerobic co-digestion of food waste and brown water. Waste Management. 2017;**61**:484-93. DOI: 10.1016/j.wasman.2016.12.013

[44] Reungsang A, Zhong N, Yang Y, Sittijunda S, Xia A, Liao Q. Hydrogen from photo fermentation. In: Green Energy and Technology. Springer Verlag; 2018. p. 221-317. DOI: 10.1007/978-981-10-7677-0_7